Society of Earth Scientists Series

Series Editor

Satish C. Tripathi, Lucknow, India

The Society of Earth Scientists Series aims to publish selected conference proceedings, monographs, edited topical books/text books by leading scientists and experts in the field of geophysics, geology, atmospheric and environmental science, meteorology and oceanography as Special Publications of The Society of Earth Scientists. The objective is to highlight recent multidisciplinary scientific research and to strengthen the scientific literature related to Earth Sciences. Quality scientific contributions from all across the Globe are invited for publication under this series. Series Editor: Dr. Satish C. Tripathi

Krishna Kumar Singh ·
Chandra Shekhar Prasad Ojha
Editors

Sustainable Management of Land, Water and Pollution of Built-up Area

Springer

Editors
Krishna Kumar Singh
Department of Civil Engineering
National Institute of Technology, Kurukshetra
Kurukshetra, Haryana, India

Chandra Shekhar Prasad Ojha
Department of Civil Engineering
Indian Institute of Technology, Roorkee
Roorkee, Uttarakhand, India

ISSN 2194-9204 ISSN 2194-9212 (electronic)
Society of Earth Scientists Series
ISBN 978-3-031-56175-7 ISBN 978-3-031-56176-4 (eBook)
https://doi.org/10.1007/978-3-031-56176-4

This Springer imprint is published by the registered company Springer Nature Switzerland AG
The registered company address is: Gewerbestrasse 11, 6330 Cham, Switzerland

Paper in this product is recyclable.

Preface

The rapid growth of population and urbanization in India together with climate change, natural hazards and human activity have posed unprecedented challenges such as landslides, urban flooding, generation of tons of solid waste and deterioration of fresh water sources and environment.

This book focuses on sustainable management of land, water and pollution. Some of the chapters in book have been selected from 1st International Online Conference on Sustainable Development in Civil and Electrical Engineering (SDCEE-2021) during December 17–19, 2021, jointly organized by Departments of Civil and Electrical Engineering, NIT Kurukshetra.

The book contains a total of 18 chapters in three broad sections: land, water and pollution. In land management part, chapters deal with monitoring spatiotemporal land use and land cover dynamics; landslide susceptibility assessment; geological characterization of dam site; increasing the stability of frozen soil; seismic soil-structure response of plan irregular building on piled raft foundation; seismic site characterization and ground response analysis for the proposed bridge.

In water management part, chapters deal with flood inundation modeling; best management practice of urban flooding using rain garden; identification of critical urban drainage flooding points; macropore flow and its characteristics; groundwater vulnerability assessment; modeling of water quality index; water and disasters.

In pollution management part, chapters deal with wastewater remediation; mitigating landfill gas emissions; bioreactor landfills; environmental performance assessment of electric and gasoline two-wheelers and decentralized wastewater treatment technologies and systems.

The editors believe that this book will serve the larger segment of the society and will contribute with the latest technology which could be helpful in mitigation of various problems related to land, water and pollution.

Kurukshetra, India — Krishna Kumar Singh
Roorkee, India — Chandra Shekhar Prasad Ojha

Contents

About the Editors

Krishna Kumar Singh, B.Tech. (IIT BHU, Varanasi), M.E. (IIT Roorkee), Ph.D. (NIT, Kurukshetra) is Professor in Civil Engineering Department, NIT, Kurukshetra. He has more than 37 years of teaching experience at NIT Kurukshetra, India, University of Rwanda, Kigali, Rwanda, and University of Technology, Lae, Papua New Guinea. He was Founder Head of Department of Civil and Environmental Engineering at the University of Rwanda, Kigali, Rwanda (January 1999–December 2002). He was Dean (Estate) at NIT, Kurukshetra (November 2012–October 2015). He was Head, Civil Engineering Department at NIT, Kurukshetra (November 2020–November 2022). He was Visiting Professor at the University of Technology, Lae, Papua New Guinea (June 2009–March 2010). He has more than 100 publications including more than 35 in Sci/Scopus index international/national journals. He has two German patents issued. He has also supervised/supervising more than 10 Ph.D. students and executed more than 50 research and consultancy projects. He has co-authored one book published by ASCE. He has been recipient of Outstanding Reviewer of *Journal of Hydrologic Engineering*, ASCE, 2010. He has been recipient of the best research paper award in the International Conference on SDCEE-2021. He is Fellow of (IOE) India, Life Member of ASCE and Life Member of ISTE.

Chandra Shekhar Prasad Ojha (1962–), Ph.D. (Imperial College, London) is Institute Chair Professor since 2018 in Civil Engineering Department at IIT Roorkee. He has more than 39 years of teaching and research experience at the University of Roorkee/IIT Roorkee. He was Head of Civil Engineering Department at IIT Roorkee (April 2015–March 2019). He was Curtis Visiting Professorship at Purdue University, USA, in 2012 and Adjunct Professor at the University of Missouri, Columbia, USA, in 2018–2020. He has more than 200 international journal publications including 80 in ASCE journals. He has also supervised 57 Ph.D. students and executed more than 75 research and consultancy projects. He has co-authored/co-edited eleven books published by Oxford University (2), ASCE (3), Elsevier (1), Springer (2), Kluwer (1), IWA (1), Taylor and Francis (1). He has been a recipient of ASCE research paper awards in 2001, 2009, 2010, 2013, 2018 and 2022; ASCE State of Award in Civil Engineering in 2014 and Visiting Fellowship of ASCE in 2010. He is a recipient of Alexander Von Humboldt Fellowship from AvH Foundation, Germany; JSPS Fellowship from Japan, STINT Fellowship from Sweden, Distinguished Visiting Fellowship of Royal Academy of Engineering UK, Commonwealth Research Scholarship from Association of Commonwealth Universities. He is Fellow ASCE since 2017 and Fellow INAE since 2020.

Sustainable Land Management

Monitoring Spatiotemporal Land Use and Land Cover Dynamics: A Case Study of Kolkata Megapolis, India (1991–2021)

Rajan Dev Gupta and Md. Omar Sarif

1 Introduction

Worldwide, the population is preferring to live in urban locations due to its pulling attraction through enriched living lifestyles and standards coupled with prosperous viabilities. In 2018, 55% of total global populace lived in impervious spaces, and if this trend goes alike then this will increase to 60% by year 2030 (UN United Nations 2018). The drastic change in land use, as per IPCC report of 2019, has resulted in the deterioration of natural land which has necessitated the monitoring and planning of micro to macro landscape to measure its threat and possible mitigation management for nourishing the land capacity and restoration (IPCC Climate Change and Land 2019). In the land transformation process, an alarming alteration is seen on all over the Earth because of high urbanization, industrialization, agriculture expansion, forest logging, electricity and mining, thereby converting the land as a scarce and limited resource in each coming day (Dubovyk 2017). This land alterations have critical effects on substantial global ecosystem (Polasky et al. 2011). Therefore, optimal use of LULC (Land Use and Land Cover) information is required for planning and implementing various schemes to achieve the immense demands for individual's simple necessities as well as welfare (Deka et al. 2019).

Globally, varied studies have been carried out on LULC where foremost purpose is to understand the land dynamics and the transformation mechanism over agriculture, water body, settlements, forest and environment factors using spatiotemporal satellite

R. D. Gupta (✉)
Civil Engineering Department and Member of GIS Cell, Motilal Nehru National Institute of Technology Allahabad, Prayagraj 211004, India
e-mail: rdg@mnnit.ac.in

Md. O. Sarif
Department of Geography, and Department of Research Impacts and Outcome, Lovely Professional University, Phagwara 144411, India
e-mail: omar.28426@lpu.co.in

K. K. Singh and C. S. Prasad Ojha (eds.), *Sustainable Management of Land, Water and Pollution of Built-up Area*, Society of Earth Scientists Series,
https://doi.org/10.1007/978-3-031-56176-4_1

datasets (Lambin et al. 2003; Oliveira et al. 2014; Mallupattu et al. 2013; Rousta et al. 2018; Sarif et al. 2017). In recent times, modern techniques like GIS (Geographic Information System) and Remote sensing are widely used to quantify the spatiotemporal changes in the natural landscape because of their proficient competencies to address small to large scale analysis at various level from local to regional to nation to the globe (Long et al. 2018; Kogo et al. 2019; Matloob et al. 2021a). These studies are possible due to accessibility of remotely sensed information through satellite imageries that depicts true ground realities. These are used for representing LULC dynamics to achieve the actual landscape status. Timely and detailed LULC information offer the understanding of the spatial pattern of the landscapes along with its inducing aspects from its past history (Mengistu and Salami 2007; Srivastava and Gupta 2003; Matloob et al. 2021b). As per literature, the primary concerns of LULC analysis is to define occurrence of change, trend types and change magnitude (Lu et al. 2004; Srivastava and Gupta 2002).

India, one of the highly populous and developing nations in world, is observing extraordinary transformation in its land topography particularly in past few decades. In current past, various studies are done for the Indian megapolis and cities such as Delhi (Chakraraborty et al. 2015), Kolkata (Sharma et al. 2015), Mumbai (Samant and Subramanyan 1998), Chennai (Aithal and Ramachandra 2016), Bengaluru (Aithal and Ramachandra 2016; Govind and Ramesh 2019), and Prayagraj (Sarif and Gupta 2021a) for representing the spatial changes of land dynamics. Kolkata is one of the leading economically vibrant and industrial megapolis of India. For Kolkata megapolis, Sahana et al. (2018) explored the spatial pattern of LULC but detailed alteration in LULC classes was missing. Khan et al. (2019) carried out built up growth during 2000–2015 but change matrix of LULC classes is not provided. Sharma et al. (2015) extracted urbanization pattern during 1989–2016 but it is lacking in descriptive transition information. Due to these existing research gaps, Kolkata megapolis, the capital of West Bengal state, is selected for present research work.

Therefore, the present research work is undertaken with main objectives being: (a) Creation of multi-temporal LULC maps for two time points, namely, 1991 and 2021 using Landsat-5 TM and Landsat-8 OLI/TIRS open-source data; and (b) Generation and analysis of change matrix of LULC dynamics over a period of 30 years during 1991–2021. The summer time (March–April) is chosen for both remote sensing datasets because these time-points are best for having good spectral information and cloud free data availability. The data of these months also provides better opportunities to understand the separations among distinct LULC classes (Fonji and Taff 2014; Pathak et al. 2021; Sarif and Gupta 2019).

2 Materials and Methods

2.1 The Study Area: Kolkata Megapolis

The study area of the present work is Kolkata Megapolis, the capital of the West Bengal State. In this work, it covers an area of 1809.56 km^2 with latitudes range as 22° 20′ 52″–22° 46′ 52" N and longitude range as 88° 06′ 43″–88° 34′ 43″ E (Fig. 1). The megapolis has Hooghly river in its central part flowing from North direction to South-West (Sahana et al. 2018). The megapolis location above Mean sea level (MSL) is about 6.26 m. The megapolis climate is tropical wet and dry type. The average annual rainfall of the city is 1640 mm. The mean monthly temperature falls in the range 19–30 °C (Sharma et al. 2015). The ancillary/secondary data is used to create 24 km buffer around the Kolkata city centre. DEM (Digital Elevation Model) is used to depict elevation details and to correlate distributed natural physiology in the study area.

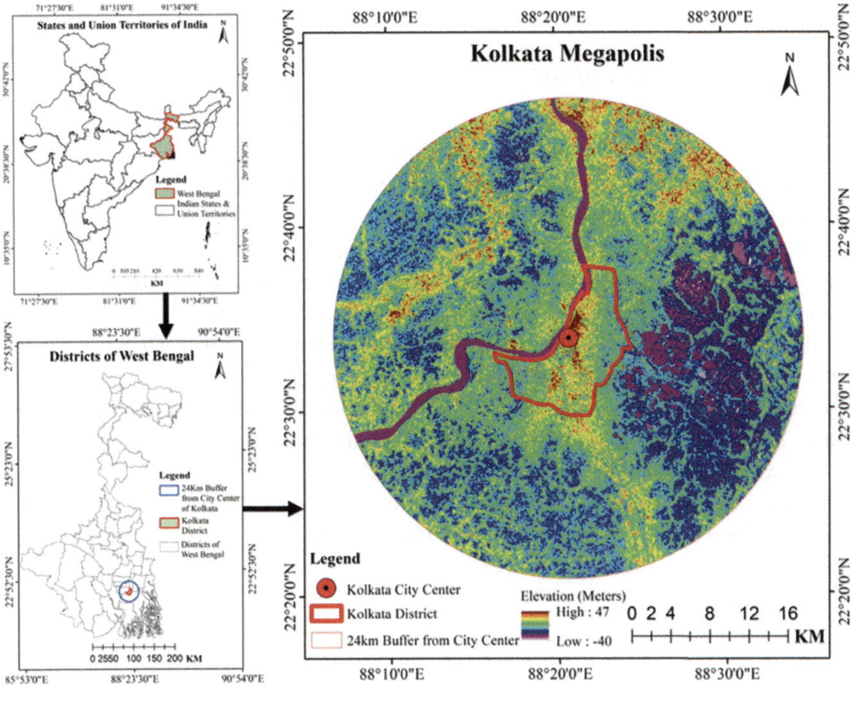

Fig. 1 Map of Kolkata Megapolis

Table 1 Satellite data used for Kolkata megapolis

Satellite (sensor)	Path/Row	Date/Time of acquisition (GMT)	Spatial resolution (m)
Landsat 5 (TM)	138/44	06-03-1991 at 03:52:09	30
	138/45	06-03-1991 at 03:52:33	
Landsat 8 (OLI/TIRS)	138/44	25-04-2021 at 04:30:36	30
SRTM DEM	–	23-09-2014	30

2.2 Data Used

In this work, Landsat Thematic Mapper (TM) digital images of 06-03-1991 and Operational Land Imager (OLI)/Thermal Infrared Sensor (TIRS) digital images of 25-04-2021 are used having spatial resolution of 30 m (Table 1). These datasets of Landsat are acquired from United States of Geological Survey (USGS) web portal (https://earthexplorer.usgs.gov/). The elevation data of Shuttle Radar Topography Mission (SRTM) is also acquired from the said USGS web portal.

2.3 Image Classification

LULC map is created by Supervised classification in which Maximum Likelihood Classifier (MLC) is used to carry out the image classification (Langat et al. 2019; Sarif and Gupta 2021b). The classification is executed after employing ground features as spectral signatures (Kogo et al. 2019; Sarif and Gupta 2021c). For performing classification of Landsat 5 (TM) image of 1991, Red band (band 3), NIR band (band 4) and SWIR band (band 5) are selected for classification. Whereas Coastal blue (band 1), NIR (band 5), SWIR 1 (band 6), SWIR 2 (band 7), Cirrus (band 9), TIR 1 (band 10) and TIR 2 (band 11) in Landsat 8 (OLI/TIRS) image of 2021 are selected as it is found to be the best band combination to perform LULC classification (Liu and Zhang 2019). Anderson's multilevel LULC classification system is used to extract the LULC classes at level-I. In the present work, the image is classified in to five LULC classes, viz., built up (BL), agriculture (AL), forest (FL), water body (WB) and barren land (BAL) (Anderson et al. 1976). Erdas Imagine 2014 image processing software is used for whole process execution.

2.4 Accuracy Assessment

Accuracy assessment is done for classified LULC images of 1991 and 2021. Total 500 points are selected on the basis of spectral signatures through random sampling method for both the classified images. After this, the validation is carried out on the basis of the knowledge of interpretation skills of authors using FCC for 1991 while FCC and Google Earth images for 2021. The same are validated using the published work of Kolkata megapolis (Sahana et al. 2018; Khan et al. 2019) also where LULC map of 1990, 2010, 2015 and 2016 were prepared. The error matrix for 1991 and 2021 is thus obtained. Then, statistical measures, i.e., user accuracy (UA), producer accuracy (PA) and overall accuracy (OA) are computed. This is followed by computation of kappa coefficient.

2.5 Change Detection and Change Matrix

Change detection of LULC between 1991 and 2021 is performed because of its significance and appropriateness in depicting nature of change (Sarif and Gupta 2021a; Langat et al. 2019). The methodology used is shown in Fig. 2. The classified maps of LULC of 1991 and 2021 are spatially subtracted for representing 30 years periodic changes in all LULC classes during 1991–2021. This approach provides the information of 'from-to' change. The change matrix (Kogo et al. 2019) or transition matrix (Rawat and Kumar 2015) approach is carried out to delineate the alteration/ conversion of one class to others and vice versa. The transition/ change matrix will give the changes, e.g., AL to BL, WB to BAL, AL to BL, BAL to BL, among others (Langat et al. 2019).

3 Results and Discussion

3.1 Multi-Temporal LULC Maps and the Classes

LULC maps of 1991 and 2021 of Kolkata Megapolis obtained by carrying out supervised classification using MLC on Landsat images are shown in Fig. 3. Further, the area covered by LULC classes in 1991–2021, both in km^2 and in percent (%) is shown in Fig. 4.

In 1991, coverage of LULC classes shows that AL was 821.40 km^2 (45.39%) which was the highest coverage area, followed by built up covering an area of 437.49 km2 (24.18%), FL as 288.00 km^2 (15.92%), WB as 184.92 km^2 (10.22%), and lastly the BAL as 77.73 km^2 (4.30%) which was the lowest. However, in 2021, area covered by built up is 1032.28 km^2 (57.05%) which is the highest. This is followed by AL covering an area as 392.27 km^2 (21.68%), FL as 148.87 km^2 (8.23%), WB as 159.04 km^2 (8.79%) and BAL as 77.10 km^2 (4.26%). In 2021 also, the BAL is again the lowest as per area coverage. In fact, similar trend in computed LULC classes is also observed in previous studies (Sahana et al. 2018; Sarif et al. 2022; Sarif and Gupta 2022).

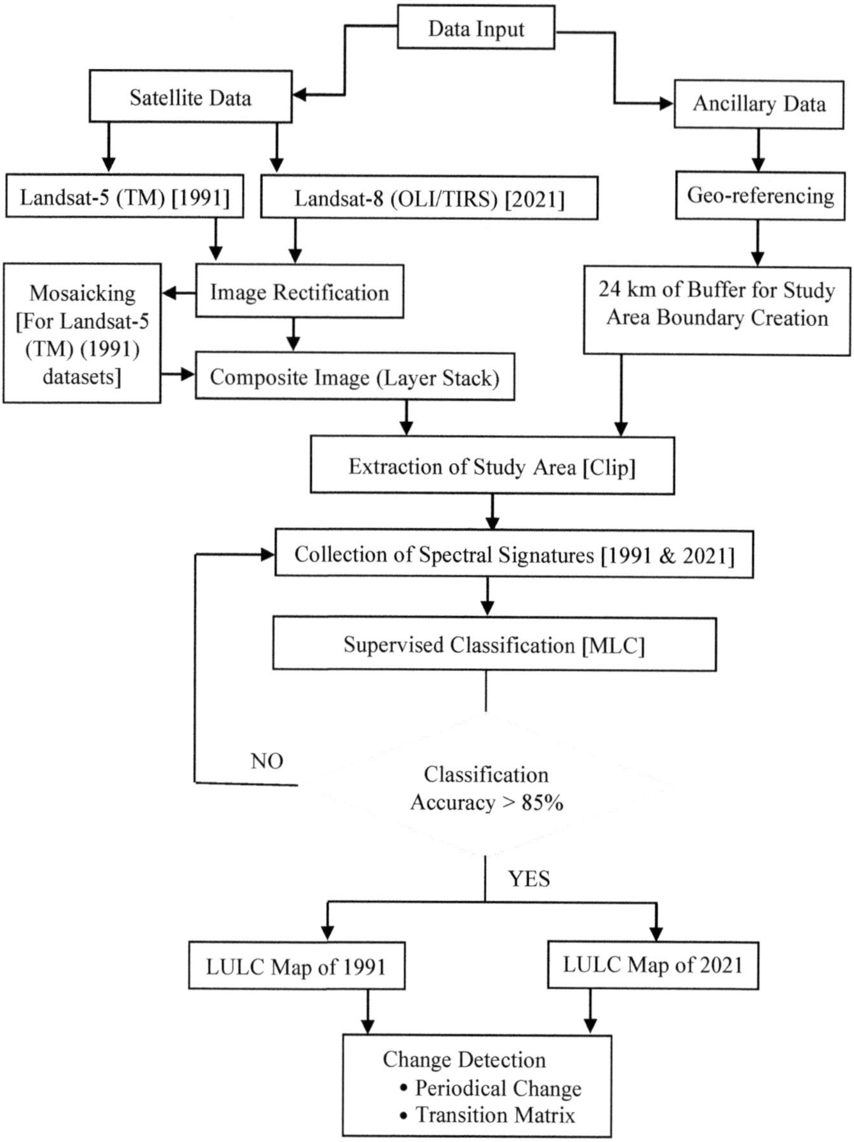

Fig. 2 Flowchart showing the methodology

3.2 *Accuracy Assessment*

The description of LULC confusion matrix for 1991 is shown in Table 2 while confusion matrix for 2021 is shown in Table 3. The user accuracy as well as producer accuracy of all LULC classes, i.e., BL, AL, FL, WB and BAL is found to be more

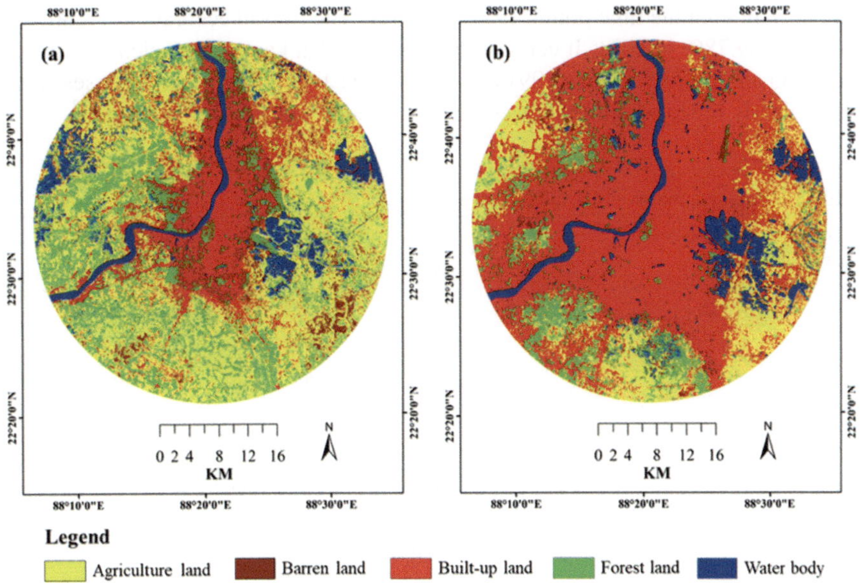

Fig. 3 LULC dynamics of Kolkata Megapolis: **a** LULC map of 1991; **b** LULC map of 2021

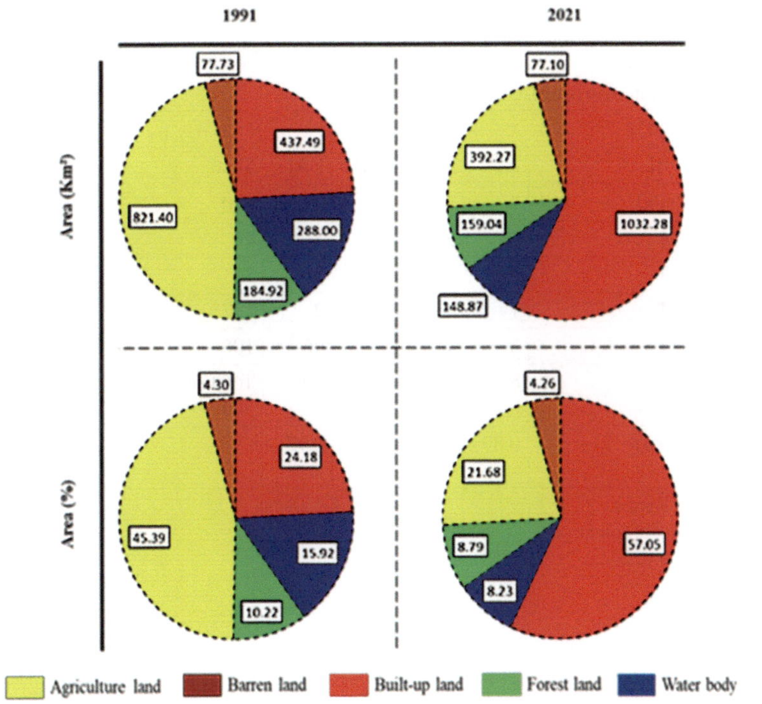

Fig. 4 Area covered by LULC classes in 1991–2021 (in km^2 and %)

than 85% in both the time points 1991 and 2021. The overall accuracy is 88.80% in 1991 and 89.78% in 2021. It can also be observed that kappa coefficient computed are 0.866 and 0.872 for years 1991 and 2021 respectively. The accuracy assessment result clearly reflects high level accuracy achieved in image classification.

Table 2 Confusion matrix of LULC map of 1991

Confusion matrix	1991						
	BL	AL	FL	WB	BAL	Total	User accuracy
BL	128	5	2	0	3	138	0.93
AL	5	96	3	2	5	111	0.86
FL	2	3	60	1	2	68	0.88
WB	3	1	4	86	2	96	0.90
BAL	7	4	1	1	74	87	0.85
Total	145	109	70	90	86	Total samples = 500	
Producer accuracy	0.88	0.88	0.86	0.96	0.86		
Overall accuracy = 88.80	Kappa coefficients = 0.866						

Table 3 Confusion matrix of LULC map of 2021

Confusion matrix	2021						
	BL	AL	FL	WB	BAL	Total	User accuracy
BL	100	3	4	1	2	110	0.91
AL	4	78	3	3	2	90	0.87
FL	3	2	68	1	2	76	0.89
WB	4	1	4	106	2	117	0.91
BAL	4	4	1	1	96	106	0.91
Total	115	88	80	112	104	Total samples = 500	
Producer accuracy	0.87	0.89	0.85	0.95	0.92		
Overall accuracy = 89.78	Kappa coefficients = 0.872						

Table 4 LULC periodical change information of Kolkata Megapolis during 1991–2021

LULC classes	Area (km^2)			Area (%)		
	1991	2021	Change	1991	2021	Change
Built up	437.49	1032.28	594.79	24.18	57.05	32.87
Forest	288.00	148.87	– 139.13	15.92	8.23	– 7.69
Water body	184.92	159.04	– 25.89	10.22	8.79	– 1.43
Agriculture	821.40	392.27	– 429.13	45.39	21.68	– 23.71
Barren land	77.73	77.10	– 0.64	4.30	4.26	– 0.04
Total	1809.56	1809.56	–	100	100	–

3.3 *Periodical LULC Dynamics*

The periodical change results indicate that all LULC classes are altered during time period from 1991 to 2021 over a period of 30 years with remarkable trend, pattern and magnitude. The growth of BL class is a major concern. The area covered by various LULC classes in 1991 and 2021 and their change dynamics is shown in Table 4. It is clear that the change that has taken place during 1991–2021 in Kolkata Megapolis is quite large and disturbing from the point of view of environmental sustainability.

During 1991–2021, the only enlarged class of LULC is built up which increased by 594.79 km^2 (32.87%) whereas the most declined class of LULC is agriculture with a decrease of 429.13 km^2 (23.71%). FL also decreased by 139.13 km^2 (7.69%) area which is substantial. Water bodies and BAL decreased marginally by 1.43% and 0.04% respectively.

It can be observed that AL is the most dominating class in 1991 with 45.39% of total land area in Kolkata Megapolis but this leading class is replaced by BL in 2021 with 57.05% of total land area. The periodical change in LULC classes is shown with the help of Fig. 5. The BL was only leading transformation class in LULC which has been intensely changed the landscape of Kolkata Megapolis in itself. The most disturbing aspect of this change is the fact that increase in built up class due to decrease of agriculture class and forest class, thereby leading to overall environmental degradation.

3.4 *Transition Matrix*

Transition/change matrix for 1991–2021 is also carried out. This is spatially shown in Fig. 6 for depicting the land transformation whereas the statistical details are compiled in km^2 areal unit in Table 5 and in percentage (%) areal unit in Table 6. This is signified in two aspects: First is from one LULC class to other class(s) of LULC and vice versa. The second is unconverted/ unchanged LULC class(s) which remain almost stagnant during said time period in landscape of megapolis.

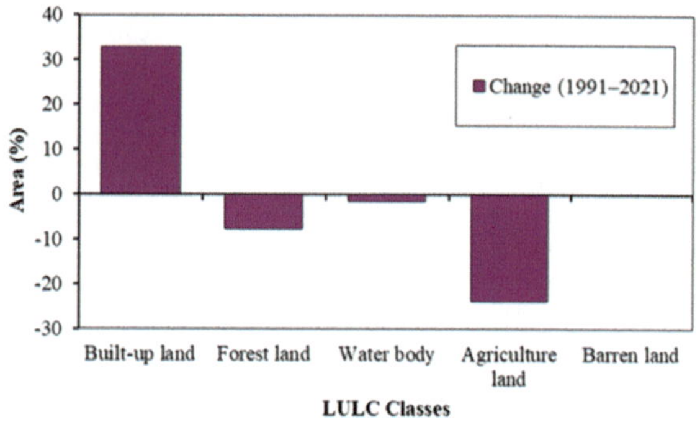

Fig. 5 LULC periodical change of Kolkata Megapolis from 1991 to 2021

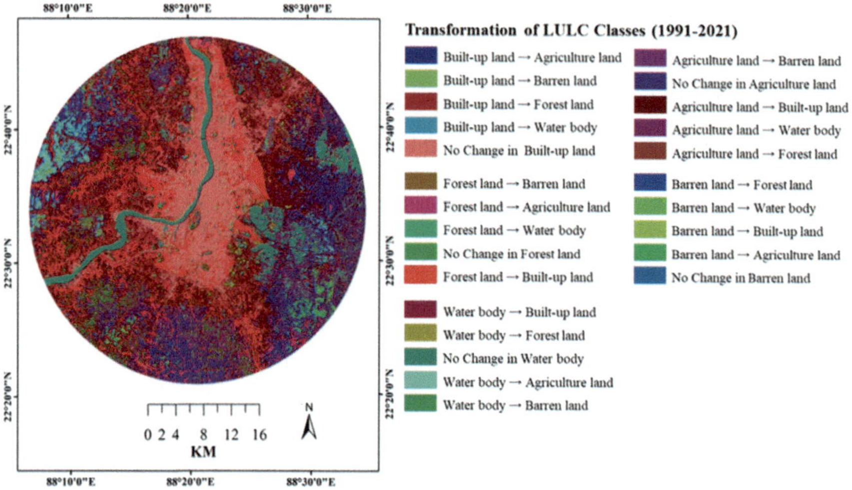

Fig. 6 The LULC transformation in Kolkata Megapolis (1991–2021)

Table 5 Change matrix (unit: km^2) of LULC classes in Kolkata Megapolis (1991–2021)

Transformation during 1991–2021 (unit: Km^2)		1991					
		BL	AL	FL	WB	BAL	Total
2021	BL	373.48	179.74	41.65	401.79	35.62	1032.28
	AL	10.75	49.36	10.51	74.90	3.35	148.87
	FL	9.79	11.39	86.94	46.20	4.72	159.04
	WB	27.69	38.20	38.15	260.85	27.38	392.27
	BAL	15.78	9.32	7.68	37.67	6.65	77.10
	Total	437.49	288.00	184.92	821.40	77.73	–

Table 6 Change matrix (unit: percentage) of LULC classes in Kolkata Megapolis (1991–2021)

Transformation during 1991–2021 (unit: %)		1991					
		BL	AL	FL	WB	BAL	Total
2021	BL	20.64	9.93	2.30	22.20	1.97	57.05
	AL	0.59	2.73	0.58	4.14	0.19	8.23
	FL	0.54	0.63	4.80	2.55	0.26	8.79
	WB	1.53	2.11	2.11	14.41	1.51	21.68
	BAL	0.87	0.51	0.42	2.08	0.37	4.26
	Total	24.18	15.92	10.22	45.39	4.30	–

During 1991–2021: From built up class, 27.69 km^2 (1.59% of total area) area changed into agriculture, 15.78 km^2 (0.87%) area changed into BAL and 10.75 km^2 (0.59%) area changed into forest. Transformation from other LULC classes to BL class is very noticeable as BL class increased due to change from AL by 401.79 km^2 (22.20%), FL by 179.74 km^2 (9.93%) and BAL by 35.62 km^2 (1.97%). The unchanged/ unconverted BL is 373.48 km^2 (20.64%) of total area of Kolkata Megapolis.

During 1991–2021: From AL, 401.79 km^2 (22.20%) area transformed into BL which is one of the major findings in this work. Whereas 74.90 km^2 (4.14%) AL area got converted into FL and 46.20 km^2 (2.55%) area converted into WB. Change from other LULC classes to AL class is very negligible as only 27.38 km^2 (1.51%) BAL area, 38.20 km^2 (2.11%) FL area and 38.15 km^2 (2.11%) WB area got converted to AL. The unchanged AL is 260.85 km^2 (14.41%) of total area of Megapolis landscape.

Between 1991 and 2021: From FL class, 179.74 km^2 (9.93%) area transformed into BL, 9.32 km^2 (0.51%) area transformed into BAL and the rest is negligible. Change from other classes of LULC to FL is very noticeable as it is transformed from AL by 74.90 km^2 (4.14%), BL by 10.75 km^2 (0.59%) and rest are negligible. Further, unchanged FL class is 49.36 km^2 (2.73%) of total area of Megapolis landscape.

Between 1991 and 2021: From WB, 41.75 km^2 (2.30%) area got transformed into BL while the rest are negligible. From other LULC classes to WB, change is high only in AL class by 46.20 km^2 (2.55%) and rest are negligible. Further, unchanged WB is 86.94 km^2 (4.80%) of total area of Kolkata Megapolis landscape.

Between 1991 and 2021: From BAL, 35.62 km^2 (3.86%) area changed into BL, 27.38 km^2 (1.51%) area got converted into AL, 4.72 km^2 (0.37%) area transformed into WB and rest is negligible. Change from other LULC classes to BAL is very evident only from AL class by 37.67 km^2 (2.08%) and rest are negligible. The unchanged BAL is 6.65 km^2 (0.37%) of total area of Megapolis landscape.

3.5 Evaluation of LULC Dynamics in Urban planning and Sustainability

During 1991–2021, in time interval of 30 years, Kolkata has witnessed an intense change in the megapolis landscape. The most declined class is AL in this urbanisation process because BL land in the city is increased by 401.79 km^2 (22.20%) on the expense of AL. The other declined class is FL with a decrease of 179.74 km^2 (9.93%) which got transformed to BL. The BAL space got decreased by 27.38 km^2 (1.51%) at the cost of agriculture. The WB has remarkably declined by 41.65 km^2 (2.30%) but it witnessed new man-made water bodies on the agriculture class by 46.20 km^2 (2.55%).

The change in FL, WB and AL has adverse consequences on the economy of the rural areas (markets, farming, grazing, etc.) as well as land functioning because of the impervious growth in the Megapolis (Sarif et al. 2022). This may adversely affect the food supply, agriculture production and access of food itself which will directly create pressure on the natural resource sustainability and food security (Langat et al. 2019; Sarif and Gupta 2022). Due to urban spatial progress, main alarming effect is intensification of Land Surface Temperature (LST) which will have critical effects on human beings and also the animals (Rousta et al. 2018).

At present stance, LULC transition/ alteration is a vital parameter contributing vital roles to extreme change in megapolis which is of critical concern to environmentalist, policy makers, scientific communities, and urban planner to mitigate high adverse effects due to landscape transition on climate change (Sarif et al. 2023). In this regard, IPCC has stated that for minimizing adverse effects of landscape change, landscape monitoring and planning on small to large scale is acutely required to achieve and improve sustainability of the environment as well as land capacity (IPCC Climate Change and Land 2019).

4 Conclusion

This study integrates geospatial methods for exploring long-term LULC dynamics over the landscape of the Kolkata Megapolis to bring out the landscape status by exploring where and how the land has transformed. The overall classification accuracy achieved for LULC maps of 1991 and 2021, generated using Landsat-5 TM and Landsat-8 OLI/TIRS images, is more than 85% for both the classified images which endorses the classification effectiveness. On the basis of the analysis of results of long-term change dynamics for a period of 30 years (1991–2021), intense development of built up (24.18–57.05%) is found. In this massive conversion, the key affected classes are agriculture (45.39–21.68%) and forest (15.92–8.23%). Further, the transition matrix results reveal the decrease of agriculture by 429.13 km^2 (23.71%) and forest by 139.13 km^2 (7.79%) resulting in impervious development in the form of built up. The water body also declined at the cost of built up by 41.65 km^2 (2.30%).

This all together aided in filling the substantial LULC transformation information gap for the Kolkata Megapolis.

The assessed loss of forest land, agriculture land and water bodies is quite large resulting in severe degradation of land, environment and quality of life as well as impacting the local climate adversely. The adopted approach in this work establishes high significance of freely available spatiotemporal digital images of Landsat satellite in LULC mapping and monitoring the landscape. The present work will add to the existing knowledge bank of LULC dynamic forces of Kolkata Megapolis and will prove to be beneficial in policy making and developmental planning to mitigate the contemporary antagonistic condition of the landscape leading to improvement in the sustainability of the environment and carrying capacity of the Kolkata Megapolis.

References

Aithal BH, Ramachandra TV (2016) Visualization of urban growth pattern in Chennai using geoinformatics and spatial metrics. J Indian Soc Remote Sens 44:617–633. https://doi.org/10.1007/s12524-015-0482-0

Anderson JR, Hardy EE, Roach JT, Witmer RE (1976) A land use and land cover classification system for use with remote sensor data. In: A revision of the land use classification system as presented in U.S. Geological Survey Circular 671, Washington

Chakraraborty SD, Kant Y, Mitra D (2015) Assessment of land surface temperature and heat fluxes over Delhi using remote sensing data. J Environ Manage 148:143–152. https://doi.org/10.1016/j.jenvman.2013.11.034

Deka J, Tripathi OP, Khan ML, Srivastava VK (2019) Study on land-use and land-cover change dynamics in Eastern Arunanchal Pradesh, N.E. India using remote sensing and GIS. Trop Ecol. https://doi.org/10.1007/s42965-019-00022-3

De Oliveira JC, Epiphanio JC, Rennó CD (2014) Window regression: a spatial-temporal analysis to estimate pixels classified as low-quality in MODIS NDVI time series. 3123–3142. https://doi.org/10.3390/rs6043123

Dubovyk O (2017) The role of remote sensing in land degradation assessments: opportunities and challenges. Eur J Remote Sens 50:601–613. https://doi.org/10.1080/22797254.2017.1378926

Fonji SF, Taff GN (2014) Using satellite data to monitor land-use land-cover change in north-eastern Latvia. Springerplus 3:1–15. https://doi.org/10.1186/2193-1801-3-61

Govind NR, Ramesh H (2019) The impact of spatiotemporal patterns of land use land cover and land surface temperature on an urban cool island: a case study of Bengaluru. Environ Monit Assess 191:1–20. https://doi.org/10.1007/s10661-019-7440-1

IPCC Climate Change and Land. An IPCC special report on climate change, desertification, land degradation, sustainable land management, food security, and greenhouse gas fluxes in terrestrial ecosystems. Summary for Policymakers (2019)

Khan A, Chatterjee S, Akbari H, Bhatti SS, Dinda A, Mitra C, Hong H, Van DQ (2019) Step-wise land-class elimination approach for extracting mixed-type built-up areas of Kolkata megacity. Geocarto Int 34:504–527. https://doi.org/10.1080/10106049.2017.1408704

Kogo BK, Kumar L, Koech R (2019) Analysis of spatio-temporal dynamics of land use and cover changes in Western Kenya. Geocarto Int 1–16. https://doi.org/10.1080/10106049.2019.1608594

Lambin EF, Geist HJ, Lepers E (2003) Dynamics of land-use and land-cover change in tropical regions. Annu Rev Environ Resour 28:205–241. https://doi.org/10.1146/annurev.energy.28.050302.105459

Langat PK, Kumar L, Koech R, Ghosh MK (2019) Monitoring of land use/land -cover dynamics using remote sensing: a case of Tana River Basin, Kenya. Geocarto Int 1–20. https://doi.org/10.1080/10106049.2019.1655798

Liu H, Zhang Y (2019) Selection of landsat8 image classification bands based on MLC–RFE. J Indian Soc Remote Sens 47:439–446. https://doi.org/10.1007/s12524-018-0932-6

Long JS, Hu C, Wang M (2018) Long-term spatiotemporal variability of southwest Florida whiting events from Modis observations long-term spatiotemporal variability of southwest Florida whiting events from MODIS Observations. Int J Remote Sens 39:906–923. https://doi.org/10.1080/01431161.2017.1392637

Lu D, Mausel P, Brondízio E, Moran E (2004) Change detection techniques. Int J Remote Sens 25:2365–2407. https://doi.org/10.1080/0143116031000139863

Mallupattu PK, Reddy J, Reddy S (2013) Analysis of land use/land cover changes using remote sensing data and GIS at an urban area, Tirupati India. Sci World J 2013:1–7. https://doi.org/10.1155/2013/268623

Matloob A, Sarif MO, Um J-S (2021a) Exploring correlation between OCO-2 XCO2 and DMSP/OLS nightlight imagery signature in four selected locations in India. Spat Inf Res 29:123–135. https://doi.org/10.1007/s41324-021-00381-x

Matloob A, Sarif MO, Um JS (2021b) Evaluating the inter-relationship between OCO-2 XCO2 and MODIS-LST in an industrial belt located at Western Bengaluru City of India. Spat Inf Res 29:257–265. https://doi.org/10.1007/s41324-021-00396-4

Mengistu DA, Salami AT (2007) Application of remote sensing and GIS inland use/land cover mapping and change detection in a part of south western Nigeria African. J Environ Sci Technol 1:99–109

Pathak C, Chandra S, Maurya G, Rathore A, Sarif MO, Gupta RD (2021) The effects of land indices on thermal state in surface urban heat island formation: a case study on Agra city in India using remote sensing data (1992–2019). Earth Syst Environ 5:135–154. https://doi.org/10.1007/s41748-020-00172-8

Polasky S, Nelson E, Pennington D, Johnson KA (2011) The impact of land-use change on ecosystem services, biodiversity and returns to landowners: a case study in the state of minnesota. Environ Resour Econ 48:219–242. https://doi.org/10.1007/s10640-010-9407-0

Rawat JS, Kumar M (2015) Monitoring land use/cover change using remote sensing and GIS techniques: a case study of Hawalbagh Block, District Almora, Uttarakhand, India Egypt. J Remote Sens Sp Sci 18:77–84. https://doi.org/10.1016/j.ejrs.2015.02.002

Rousta I, Sarif MO, Gupta RD, Olafsson H, Ranagalage M, Murayama Y, Zhang H, Mushore TD (2018) Spatiotemporal analysis of land use/land cover and its effects on surface urban heat island using landsat data: a case study of Metropolitan City Tehran (1988–2018). Sustainability 10:1–25. https://doi.org/10.3390/su10124433

Sahana M, Hong H, Sajjad H (2018) Analyzing urban spatial patterns and trend of urban growth using urban sprawl matrix: a study on Kolkata urban agglomeration India. Sci Total Environ 628–629:1557–1566. https://doi.org/10.1016/j.scitotenv.2018.02.170

Samant HP, Subramanyan V (1998) Landuse/land cover change in Mumbai-Navi Mumbai cities and its effects on the drainage basins and channels-a study using GIS. J Indian Soc Remote Sens 26:1–6. https://doi.org/10.1007/BF03007333

Sarif MO, Gupta RD (2021b) Comparative evaluation between Shannon's entropy and spatial metrics in exploring the spatiotemporal dynamics of urban morphology: a case study of Prayagraj City, India (1988–2018). Spat Inf Res 29:961–979. https://doi.org/10.1007/s41324-021-00406-5

Sarif MO, Gupta RD (2021c) Modelling of trajectories in urban sprawl types and their dynamics (1988–2018): a case study of Prayagraj City (India). Arab J Geosci 14:1–21. https://doi.org/10.1007/s12517-021-07573-7

Sarif MO, Gupta RD (2022) Evaluation of seasonal ecological vulnerability using LULC and thermal state dynamics using landsat and MODIS data: a case study of Prayagraj City, India

(1987–2018). Environ Sci Pollut Res 29:77502–77535. https://doi.org/10.1007/s11356-022-21225-7

Sarif MO, Gupta RD, Murayama Y (2023) Assessing local climate change by spatiotemporal seasonal LST and six land indices, and their interrelationships with SUHI and hot-spot dynamics: a case study of Prayagraj City, India (1987–2018). Remote Sens 15:1–40. https://doi.org/10.3390/rs15010179

Sarif MO, Gupta RD (2019) Land surface temperature profiling and its relationships with land indices: a case study on Lucknow City. In: Proceedings of the ISPRS annals of photogrammetry, remote sensing and spatial information sciences, vol IV-5/W2. Copernicus Publications, Dhulikhel, Nepal LAND, Dec 10 2019, pp 89–96

Sarif MO, Gupta RD (2021) Spatiotemporal mapping of land use/land cover dynamics using remote sensing and gis approach: a case study of Prayagraj City, India (1988–2018). Environ Dev Sustain 1–33. https://doi.org/10.1007/s10668-021-01475-0

Sarif MO, Jeganathan C, Mondal S (2017) MODIS-VCF based forest change analysis in the State of Jharkhand. Proc Natl Acad Sci India Sect A Phys Sci 87:751–767. https://doi.org/10.1007/s40010-017-0446-6

Sarif MO, Ranagalage M, Gupta RD, Murayama Y (2022) Monitoring urbanization induced surface urban cool island formation in a south Asian megacity: a case study of Bengaluru, India (1989–2019). Front Ecol Evol 10. https://doi.org/10.3389/fevo.2022.901156

Sharma R, Chakraborty A, Joshi PK (2015) Geospatial quantification and analysis of environmental changes in urbanizing City of Kolkata (India). Environ Monit Assess 187:1–12. https://doi.org/10.1007/s10661-014-4206-7

Srivastava SKm Gupta RD (2002) Urban sprawl mapping using digital image processing techniques: a case study of Allahabad City. In: Proceedings of the ISPRS TC VII international symposium on resource and environmental monitoring, Hyderabad

Srivastava SK, Gupta RD (2003) Monitoring of changes in land use/land cover using multi-sensor satellite data. In: Proceedings of the 6th international conference on GIS/GPS/RS: Map India 2003, New Delhi

UN United Nations, Department of Economic and Social Affairs, Population Division (2018). World urbanization prospects: The 2018 Revision, Online Edition. 2018

Landslide Susceptibility Assessment Using Mathematical Modeling in GIS for Uttarakhand, India

Lakhwinder Singh, Chandra Shekhar Prasad Ojha, and Deepak Khare

1 Introduction

Landslide refers to the sliding of an upper layer of soil. This usually happens in mountain regions with steep slope areas. There are a number of factors that are responsible for landslide event and may include precipitation, elevation, slope and manmade constructions, etc. Landslides cause risk to humans, animals, settlements, roads and natural vegetation. While driving in Himalayas, people generally see a warning sign board of 'landslide prone area', or 'drive carefully', 'stone may fall at any time'. So, study of landslide hazard area becomes essential. Generally. landslide study is performed on small study areas such as Uttarkashi, Pithoragarh and Dehradun (Kumar 2017; Tran et al. 2021; Pankaj and Kumar 2009; Pham et al. 2017a). In the present study, a larger region is selected for analysis, covering about 53 thousand square kilometers.

Landscape development is a complex process influenced by many natural and human-made causes at various scales, from the local to the regional (Pilz et al. 2021). Weathering and eroding geomorphological processes have a major impact on the mountain terrain direct and indirect natural features such as natural flow of water and precipitation or losing the strength of rock by chemical weathering (Dudeja et al. 2021; Ansari et al. 2021). Human activities such as road building, extraction of minerals, population growth, and deforestation are among the most common human influences that modify the landscape rapidly. To better understand a landscape's dynamic behavior, physical surface factors must be examined (Kumar

L. Singh (✉) · D. Khare
Department of Water Resources Development and Management, Indian Institute of Technology, Roorkee 247667, India
e-mail: lsingh@wr.iitr.ac.in

C. S. P. Ojha
Civil Engineering Department, Indian Institute of Technology, Roorkee 247667, India

K. K. Singh and C. S. Prasad Ojha (eds.), *Sustainable Management of Land, Water and Pollution of Built-up Area*, Society of Earth Scientists Series,
https://doi.org/10.1007/978-3-031-56176-4_2

and Anbalagan 2016; Satterthwaite et al. 2007). Landslides in the Himalayas may have devastating effects on the lives of the people who live there, their environment, and their socioeconomic situation. Because of the region's unique geotectonic, the Himalayan mountains are notorious for triggering several landslides each year during the monsoon season (Pankaj and Kumar 2009; Chaturvedi et al. 2017; Gupta et al. 2021). Increased industrial activity and development in landslide-prone areas have exacerbated the steeper slopes and deep weathering of bedrock due to throw waste water in rivers. According to the Geological Survey of India, about 0.14 million square kilometers of Uttarakhand, Himachal Pradesh, and Jammu and Kashmir are at risk of landslides (Sur et al. 2021). In the last decade, Uttarakhand state's Lesser Himalayas have seen a significant increase in development activity (Agarwal et al. 2022). Increases in natural disasters and severe climatic conditions have occurred from human activity. The most common causes of landslides are severe rains, earthquakes, or anthropogenic factors such as driving on unstable roads. According to previous research assessments, landslides frequently occur in India's Himalayan area because of a combination of hydro-geological complexity and increasing anthropogenic activity (Zeiger and Hubbart 2016; Shukla and Gedam 2019). In hilly terrain, changes in land use and land cover can alter the likelihood of a landscape. Studies show that the frequency of landslides rises in correlation with the extent to which the terrain is altered (Gupta et al. in preparation; Pham et al. 2020). Landslides may be linked to physical and hydro-meteorological factors, or they may be the result of an area's growing economic activity, as shown by the historical correlation of changes in terrain (Pham et al. 2017a; Sharma and Mahajan 2018). Many hydro-geomorphological processes as well as human activities exacerbate the dislocation of soils on steep slopes, which is the primary cause of landslides (Pham et al. 2020). Soil is well stable in places with more forest and plant cover than in areas where extensive development activities occur, such as deforestation and construction on hill slopes. Climate change can have a significant influence on the tropical belt's steep topography, such as the Lesser Himalayas (Tran et al. 2021; Pankaj and Kumar 2009; Pham et al. 2017b). It's possible that this might lead to erratic precipitation, which could enhance the danger of landslides by several times, posing a serious threat to people's lives and damaging infrastructure. According to recent research, human activity may frequently modify the natural terrain leading to complicated hydro geomorphological processes culminating in landslides (Singh and Khare 2021; Sajikumar and Remya 2015; Tian et al. 2014; Prasad and Ramesh 2018; Mehdi et al. 2015; Marhaento et al. 2018). However, landslides' effect on the natural environment and ecology has received far less attention than their impact on people and infrastructure. Studies have also shown that the Himalayan region's delicate terrain is particularly vulnerable to natural disasters, raising concerns among climate scientists about both the present and the future (Dudeja et al. 2021; Bopche and Rege 2021a; Lwarkha et al. 2020). Physiography and climate change have also been identified as important factors contributing to the worsening of mountain landscape alterations. The upshot is that certain younger, formerly non-anthropogenic landscapes may suffer landslides in diverse forms where the direct and indirect impacts of landslides may occasionally endure throughout time. Damages and losses caused by a landslide may be traced back to the spot where it occurred. In contrast, indirect costs include environmental repercussions, livelihood changes, changes in farming techniques, transportation disruptions, economic constraints, and so forth. Seismically active mountainous and steep terrain is heavily fractured and dilapidated due to the strong folds and faults in the Himalayas that is responsible of falling stone due to lower scale earthquakes. More and more parts of the Lesser Himalayas are becoming focal locations for increasing economic activity due to human involvement in ongoing building, deforestation, inappropriate agriculture techniques, increased traffic movements and mining.

The Himalayas have more frequent landslides due to all of these elements combined with hydro-geological and climatological variables (Dudeja et al. 2021; Bopche and Rege 2021b). As a result, it is critical to conduct a thorough investigation of landslide occurrences in these locations in order to implement appropriate fortification and mitigation measures to decrease landslide dangers and lessen the landscape's vulnerability. Other aspects of landslide prediction are being addressed by a variety of advanced methodologies and approaches that are being used in different regions of the world. The accuracy and predictive power of certain strategies can vary greatly.

Landslide susceptibility evaluation based on qualitative and quantitative methodologies has been the subject of a number of studies in the last decade (Tran et al. 2021; Sur et al. 2021; Kainthura and Sharma 2021). Landslide susceptibility index mapping is a common and fundamental method for estimating landslide danger. As a result of the data-driven process, it's possible to map out potential landslide zones using statistical models accurately. For decision-makers and authorities, these tools can help them make informed decisions on how to minimize the damage and losses from current and future landslides. Landslide susceptibility was mapped using a GIS-based multi-criteria decision-making technique, such as a fuzzy numerical risk factor model.

Researchers used a number of techniques to identify the possible location of landslide locations. These numerical methods may be varying depending on the quality of input data. There is a research gap for larger region analysis using proper mathematical modelling using advanced computing for landslide analysis (Pham et al. 2015). This study investigates and identifies the vulnerable landslide zones by combining mathematical models with GIS and derives landslide indices.

2 Study Area and Data Used

2.1 Study Area

This study is performed on a Himalayan state named Uttarakhand, which lies in North India, with about 53 thousand kilometers square. The minimum elevation of the region is 120 m and the maximum elevation is 7114 m above MSL (Mean Sea Level). Uttarakhand has about 2000 km of National Highways and 4500 km of state

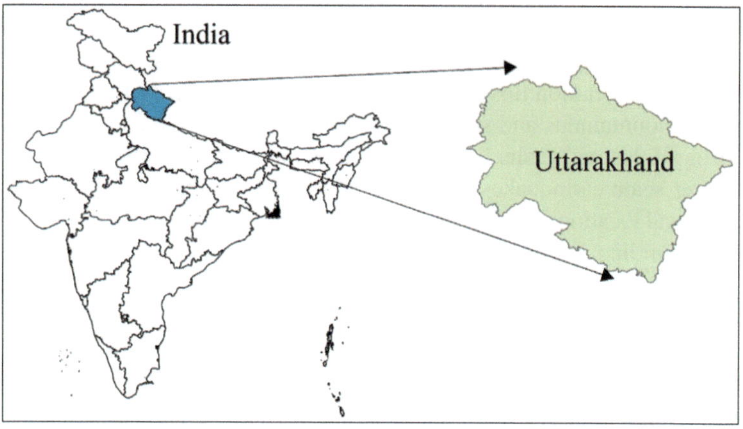

Fig. 1 Location map of study area

highways PWD: About Us: Public Works Department, Uttarakhand (n.d.). The detail of the study area is presented in Fig. 1.

2.2 Data Used

The landslide susceptibility analysis required physical and weather data. Carto-SAT based Digital elevation model of 30 m resolution is obtained from the Bhuvan portal of ISRO. Precipitation data is of 0.5° resolution used from 2010 to 2020 obtained from Climatic Research Unit of University of East Anglia, more detail about this data can be seen in Harris et al. (2020). Open street map is used to extract significant road information. Historical landslide event data is obtained from a report published by Govt. of Uttarakhand (Joshi and Burkitt 2019).

3 Methodology

This analysis is based on physical layers along with one road layer. These layers are used to generate factor affecting landslide. DEM is used to generate slope, elevation and drainage density. Precipitation grid data is used to generate precipitation layer. Road map is used to generate road distance layers. Landslide risk zones are created using weightage and influence method with the overlay approach. There is no fixed method available to set weightage value for layers. These values are based on the ranking of class and class interval, based on feature importance and set based on knowledge of study area. Ranking values are decided on the knowledge of study area and possible risk of higher values, such as higher the slope has more risk to landslide, and if this higher slope is near to road, it may cause more risk to road. Based on such conditions, values are decided. Similarly, more drainage density with more slope tends to more risk and get a higher rank. This influence and rank are decided on true condition of study area and knowledge of user. Higher the risk range is assigned the higher rank. Detail of decided values are shown in Table 1. Throughout the analysis, natural brake class intervals are used to represent the spatial distribution of data ignored by equal class intervals. The natural class interval works by dividing the class into such a way that more variation gets smaller class size and less variation gets larger class size by considering the distribution of data on the surface.

An equation is also developed after studying the factors of the study area. This landslide equation take input from physical factors and provide an index of risk of each pixel in selected location. Based on the maximum and minimum values of

Table 1 Influence and weightage table

Class	Influence Value	Range	Area km^2	Class ranking
Slope in degrees	30	0–9.94	11,671.3	1
		9.94–20.88	11,775.9	2
		20.88–30.16	14,984.8	3
		30.16–40.10	12,957.2	4
		40.10–84.50	5423.86	5
Elevation (m)	10	120–925	14,186.8	2
		925–1924	18,476.4	5
		1924–3156	8985.19	4
		3156–4541	6598.21	3
		4541–7184	8608.58	1
Drainage density	20	0.05–0.32	3745.78	1
		0.32–0.58	16,042.5	2
		0.58–0.84	26,195.3	5
		0.84–1.11	6546.64	1
		1.11–1.37	4324.97	1
Precipitation (mm)	30	656–852	8122.01	1
		825–990	10,522.3	2
		990–1110	18,200.5	3
		1110–1208	9277.93	4
		1208–1389	7689.67	5
Road distance (m)	10	0–3330	23,778.4	5
		3330–7819	18,072.7	4
		7819–13,901	9070.2	3
		13,901–22,589	4060.89	2
		22,589–36,925	1873.1	1

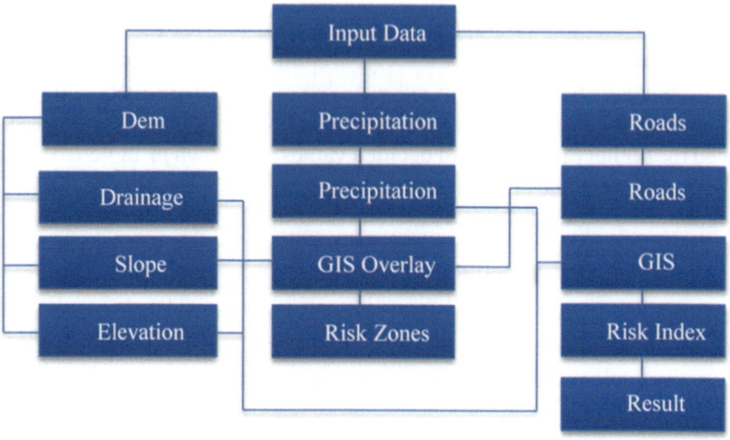

Fig. 2 Workflow of GIS process

the equation, it is mathematically converted into an index and a risk index map is generated. Later, this equation is used in the GIS model and results are derived. The output of this equation is validated with historical events of the landslide of the region. Its results were also compared with the zone indicating landslide risk map of the study area. The detailed workflow process is shown in Fig. 2. It is found that most of the historical event spatially lie in the high-risk zone of the area. The derived equation is as follows:

$$LR = Sd \times Dd \times Rf \times \begin{pmatrix} \left(4 \times 10^{-11}\right) \times E^3 - \left(5 \times 10^{-7}\right) \times E^2 \\ +0.0014 \times E + 0.63 \end{pmatrix} \quad (1)$$

where LR is Landslide risk Index, Sd is slope in degrees, Dd is length of drainage features per km^2, Rf is Mean annual precipitation and E is elevation from mean sea level.

This equation is developed in such a way such as if the degree of slope increases, then risk will increase, the same is applicable to other factors of equation. But elevation is considered in such a way that it focusses areas that have most probable to risk of landslide. This study area includes a large region and results are provided in the 30 m resolution. All layers used in this analysis are of 30 m resolution also.

4 Results and Discussion

The number of secondary layers are generated based on primary layers, such as road distance, slope and drainage density. Later, all data is combined in GIS modeling. Details are discussed in relevant sections.

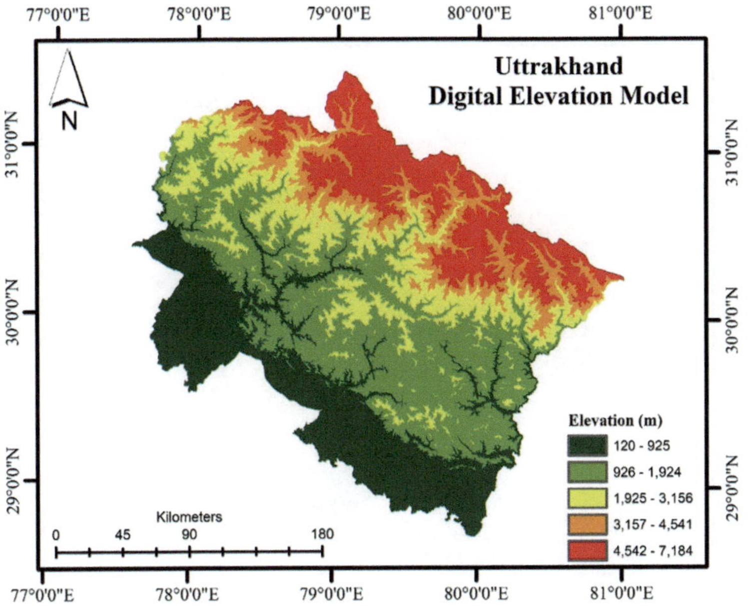

Fig. 3 Digital elevation model—Uttarakhand

4.1 Elevation

Digital elevation layer is created from Cartosat-2, version 3 data of 30 m resolution. This layer was further used to generate secondary layers. Based on historical events, most landslide events occurred in 1000–4500 m elevation (Joshi and Burkitt 2019). Higher elevation is found in the northern side of the state and lower elevation in the southern side of the state. A detailed elevation map is shown in Fig. 3.

4.2 Slope

The slope map is derived from DEM. Most of the region lies between 20° and 40° of slope, i.e. 27,942 km^2 and the mid-region of the area lies under this zone. The 5423 km^2 area is under a higher slope range of 40–84°, but it has less risk due to very few populations in this area. Higher the slope angle tends to higher the risk of landslide. A detailed slope map is shown in Fig. 4.

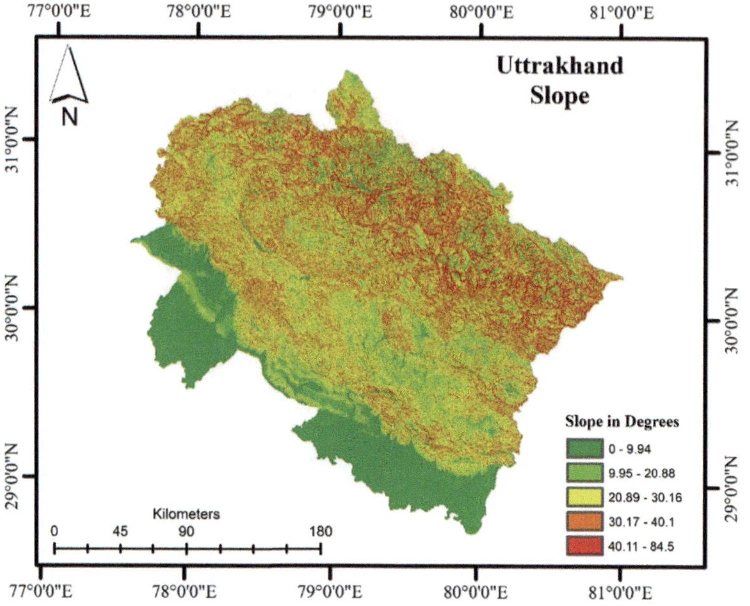

Fig. 4 Slope map of Uttarakhand

4.3 Drainage Density

Drainage density is counted by counting the total length of drainage features per km^2. More drainage density represents more the water flow during a rainfall event. This makes soil unstable, especially on steep slopes. Higher drainage density tends to more risk of landslide. So, weightage to layer is given accordingly and used as a factor in a derived equation. Generally, most researchers skip this factor which is more significant for landslide risk zone calculation, e.g. (Tran et al. 2021). Detailed map of the landslide is shown in Fig. 5.

4.4 Precipitation

The annual average precipitation pattern is generated using data from 2010 to 2020. Precipitation is calculated for each pixel, and a pattern map is drawn for the whole region from 10 years of data. Thus, this area has a much higher elevation more than 4000 m, and higher altitudes receive less precipitation. Higher elevation has the lowest mean precipitation as 656–851 mm. In contrast, lower and mid-range elevation from 120 to 3000 m receive most of the precipitation 989 m to 1389 mm. Then these become more vulnerable to landslides, including higher drainage density

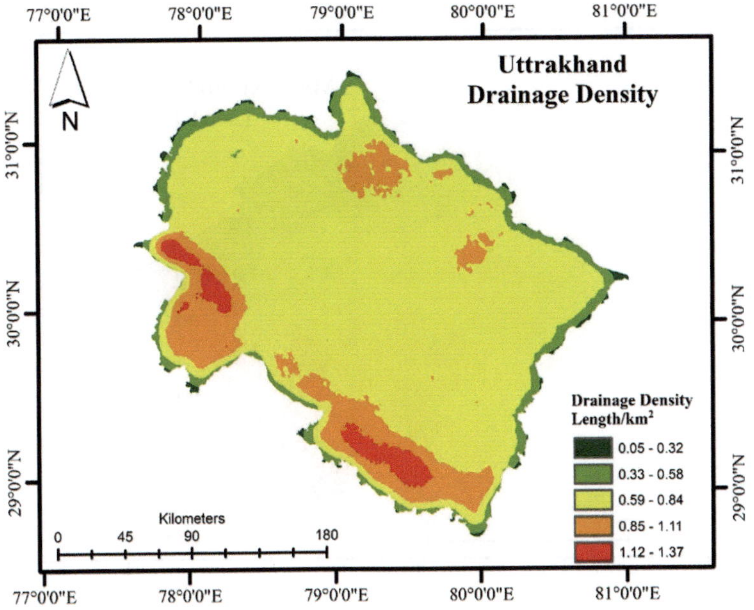

Fig. 5 Drainage density—Uttarakhand

and steep slopes. So, weightage is given accordingly, as already shown in Table 1. A detailed precipitation pattern map is shown in Fig. 6.

4.5 Road Distance

Roads are manmade features on the earth's surface. Roads are highly impacted by landslides. On other side, roads also cause the land to slide on steep slopes. Landslide events also impact settlements along roads. The number of natural drains crosses from the roads that cause landslides during a rainfall event. Roads highly impact and get impacted in this region. Even roads are mandatory for the Himalayan region to contribute to the general supply chain to hilly cities. During the construction of roads, loosening of mountain soil takes place and make it unstable. So, distance from roads matters. A location on steep slope and near to road with high drainage density will be at more risk. Considering these facts, weightage is given to this layer. Detailed road distance map is shown in Fig. 7

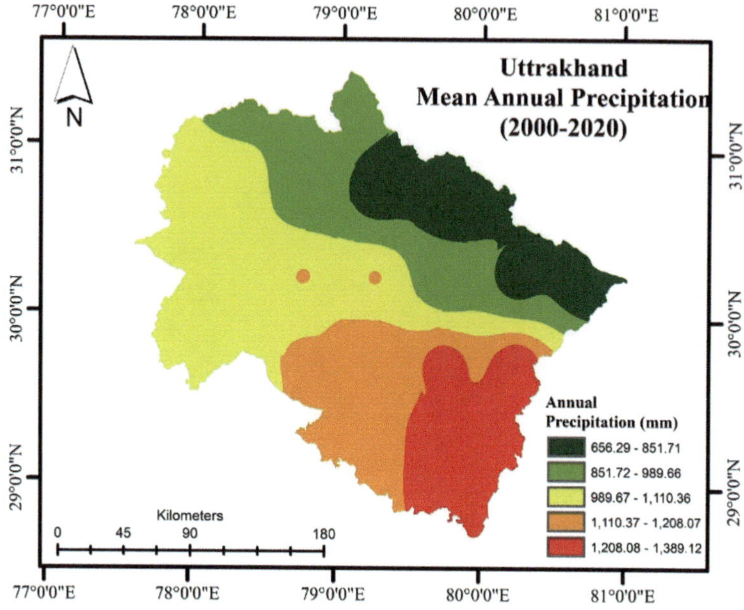

Fig. 6 Mean annual precipitation pattern—Uttarakhand

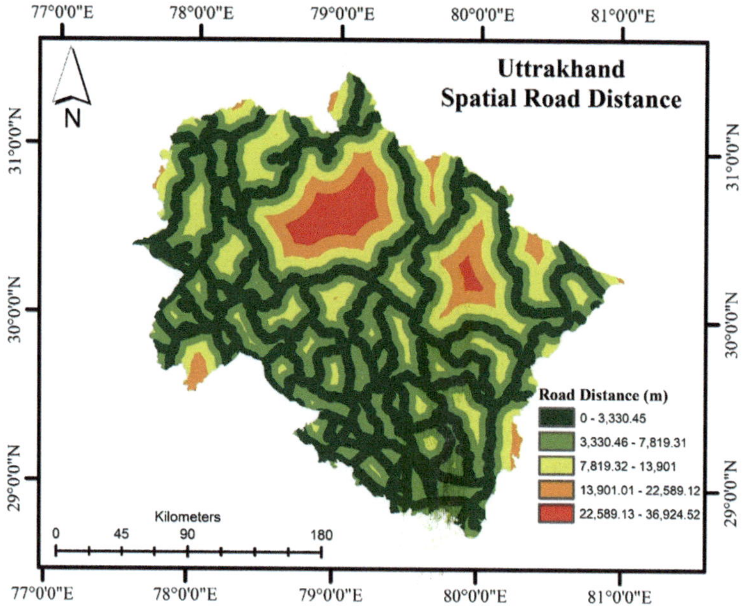

Fig. 7 Road distance map—Uttarakhand

Table 2 Landslide susceptibility zones, index and historical landslide events

	Index range		Equation based	Weighted based	Historical events
Landslide risk	Min	Max	Area km^2	Area km^2	Count
Safe	0.000	0.111	465.935	559.367	1
Least	0.111	0.178	12,101.6	12,392.6	332
Average	0.178	0.373	23,254.7	23,097.9	2645
High to Medium	0.373	0.575	16,709.9	16,514.7	3256
High	0.575	1.000	1240.54	1207.99	302

4.6 Landslide Susceptibility Assessment and Validation

Landslide susceptibility analysis was performed by considering all possible physical GIS layers. Two outputs from the same data are derived using two approaches, Mathematical GIS model-based and weightage-based zones. Results are summed up into five risk zones, and the risk index is generated between 0 and 1. Where 0 represent lowers risk or safe area and 1 represents higher risk. An index value of 0.57–1 is considered as higher risk area, and 0.37–0.57 index range represent area under high to medium risk. A high-risk zone represents more the magnitude of a landslide because of the steep slope, high rain and more drainage density. In the high-risk zone, 302 events of landslides are recorded. Most of Uttarakhand lies in average to medium risk of landslide, covering 23,097 $km^2$and 16,514 km^2 this area has recorded the highest number of historical landslide events, i.e. 2645 and 3256. This includes high to medium magnitude of landslides. On the other side, the safe and most minor risk area is just 559 km^2 and 12,392 km^2. While comparing derived risk zone results with actual historical events, most events are validated that lie spatially in average to high-risk zone and follow the same pattern as derived. Detailed results, including index range and historical events, is shown in Table 2 and Figs. 8, 9 and 10

5 Conclusions

This study demonstrated the use of GIS to identify the Landslide vulnerability zones for the larger region. Generally, using the traditional non-GIS survey approach, this analysis was much typical to perform. Physical layers of area are considered for analysis because these layers are not changed by a short time. A landslide equation is also derived to generate a landslide risk index. The proposed equation applies to the whole Himalayas because results are derived from a larger region covering lower to the higher elevation in the Himalayas. All the factors that cause landslides to happen are considered. The safe limit and risk limit of each factor is considered. The steep slope has more risk of landslide events happening. Risk will increase if the area

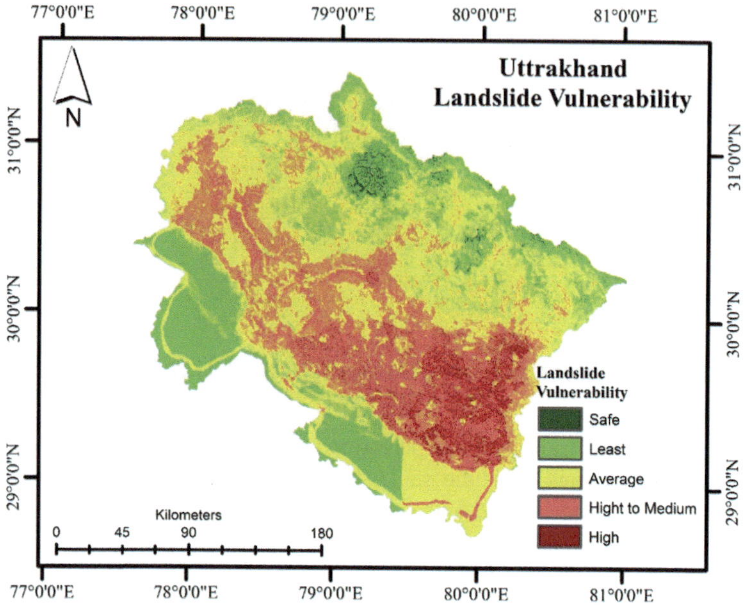

Fig. 8 Landslide vulnerability zones

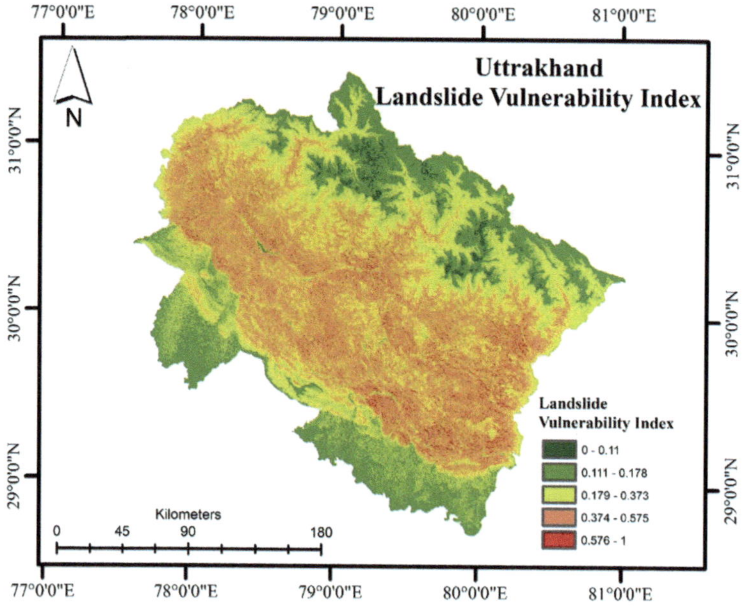

Fig. 9 Landslide risk index

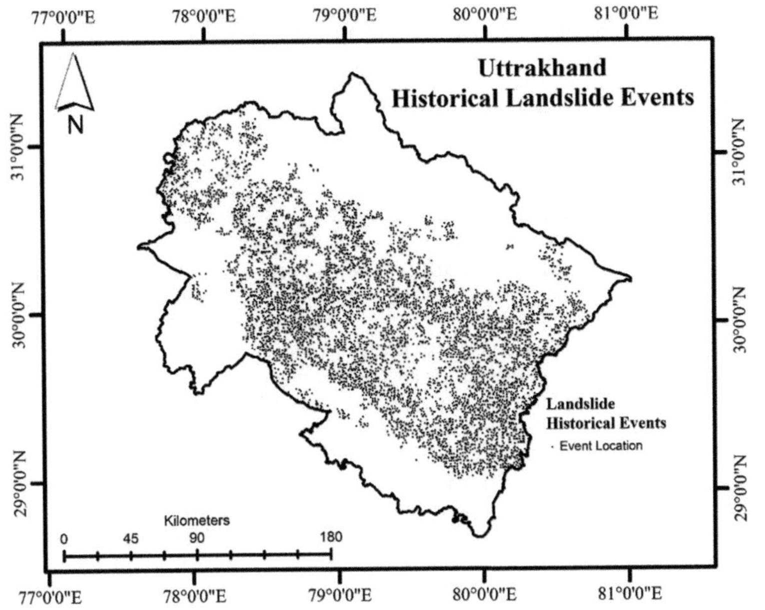

Fig. 10 Historical landslide events. *Source* (Joshi and Burkitt 2019)

has more slope, more drainage density and more rainfall. All factors are carefully considered, and the equation is derived. The derived equation will be directly helpful for the user by generation factors only. This study will help to identify the zones to apply safety measures.

References

Agarwal S, Kumar V, Kumar S, Sundriyal Y, Bagri DS, Chauhan N, Kaushik S, Khan F, Rana N (2022) Identifying potential hotspots of land use/land cover change in the last 3 decades, Uttarakhand, NW Himalaya. https://doi.org/10.31223/X5VK9F

Ansari T, Kainthola A, Singh KH, Singh TN, Sazid M (2021) Geotechnical and micro-structural characteristics of phyllite derived soil; implications for slope stability, Lesser Himalaya, Uttarakhand, India. Catena 196. https://doi.org/10.1016/j.catena.2020.104906

Bopche L, Rege PP (2021) Feature-based model for landslide susceptibility mapping using a multi-parametric decision-making technique and the analytic hierarchy process. Sadhana—Acad Proc Eng Sci 46. https://doi.org/10.1007/s12046-021-01648-7

Bopche L, Rege PP (2021b) Feature-based model for landslide prediction using remote sensing and digital elevation data. Lect Notes Electr Eng 703:299–312. https://doi.org/10.1007/978-981-15-8391-9_22

Chaturvedi P, Srivastava S, Kaur PB (2017) Landslide early warning system development using statistical analysis of sensors' data at Tangni landslide, Uttarakhand, India. Springer 547:259–270. https://doi.org/10.1007/978-981-10-3325-4_26

Dudeja D, Bhatt SP, Sharma V, Biyani AK (2021) Evaluating stability of perilous slide zones occurring in Lesser Himalayan schuppen belt between Tiuni and Taluka route, Uttarakhand, India. Environ Earth Sci 80. https://doi.org/10.1007/S12665-021-09773-6

Gupta V, Paul A, Kumar S, Dash B (2021) Spatial distribution of landslides vis-à-vis epicentral distribution of earthquakes in the vicinity of the Main Central Thrust zone, Uttarakhand Himalaya. tstwww.currentscience.ac.in. 120

Gupta SK, Shukla DP, Thakur M (2018) Selection of weightages for causative factors used in preparation of landslide susceptibility zonation (LSZ). Geomatics Nat Hazards Risk 9:471–487. https://doi.org/10.1080/19475705.2018.1447027

Harris I, Osborn TJ, Jones P, Lister D (2020) Version 4 of the CRU TS monthly high-resolution gridded multivariate climate dataset. Sci Data 7:1–18. https://doi.org/10.1038/s41597-020-0453-3

Joshi GC, Burkitt T (2019) Disaster risk assessment of Uttarakhand

Kainthura P, Sharma N (2021) Probabilistic approach to predict landslide susceptibility based on dynamic parameters for Uttarkashi, Uttarakhand (India). J Sci Ind Res (india) 80:716–725

Kumar D (2017) Study and prediction of landslide in Uttarkashi, Uttarakhand, India Using GIS and ANN. Am J Neural Netw Appl 3:63. https://doi.org/10.11648/j.ajnna.20170306.12

Kumar R, Anbalagan R (2016) Landslide susceptibility mapping using analytical hierarchy process (AHP) in Tehri reservoir rim region, Uttarakhand. Springer 87:271–286

Lwarkha O, Tahsil C, District G, Khanduri S (2020) Formation and Failure of Natural Dams in Uttarakhand Himalaya : An Observation from Lwarkha, Chamba Tahsil of Tehri Garh. Int J Earth Sci Knowl Appl 3:12–22. dergipark.org.tr.

Marhaento H, Booij MJ, Hoekstra AY (2018) Hydrological response to future land-use change and climate change in a tropical catchment. Hydrol Sci J 63:1368–1385. https://doi.org/10.1080/02626667.2018.1511054

Mehdi B, Ludwig R, Lehner B (2015) Evaluating the impacts of climate change and crop land use change on streamflow, nitrates and phosphorus: a modeling study in Bavaria. J Hydrol Reg Stud 4:60–90. https://doi.org/10.1016/j.ejrh.2015.04.009

Pankaj A, Kumar P (2009) GIS-based morphometric analysis of five major sub-watersheds of Song River, Dehradun District, Uttarakhand with special reference to landslide incidences. Springer 37:157–166

Pham BT, Prakash I, Dou J, Singh SK, Trinh PT, Tran HT, Le TM, Van Phong T, Khoi DK, Shirzadi A, Bui DT (2020) A novel hybrid approach of landslide susceptibility modelling using rotation forest ensemble and different base classifiers. Geocarto Int 35:1267–1292. https://doi.org/10.1080/10106049.2018.1559885

Pham BT, Tien Bui D, Indra P, Dholakia MB (2015) Landslide susceptibility assessment at a part of Uttarakhand Himalaya, India using GIS—based Statistical Approach of Frequency Ratio Method. Int J Eng Res V4:338–344. https://doi.org/10.17577/ijertv4is110285

Pham BT, Bui DT, Dholakia MB, Prakash I, Pham HV, Mehmood K, Le HQ (2017) A novel ensemble classifier of rotation forest and Naïve Bayer for landslide susceptibility assessment at the Luc Yen district, Yen Bai Province (Viet Nam) using GIS. Geomatics Nat Hazards Risk 8:649–671. https://doi.org/10.1080/19475705.2016.1255667

Pham BT, Tien Bui D, Prakash I, Nguyen LH, Dholakia MB (2017) A comparative study of sequential minimal optimization-based support vector machines, vote feature intervals, and logistic regression in landslide susceptibility assessment using GIS. Environ Earth Sci 76. https://doi.org/10.1007/s12665-017-6689-3

Pilz M, Cotton F, Cook K, Dieze M, Hovius N, Rekapalli R (2021) Seismic observations of the 2021 Uttarakhand landslide/debris flow and flood events. In: EGU general assembly conference abstracts, p 16583

Prasad G, Ramesh MV (2018) Spatio-temporal analysis of land use/land cover changes in an ecologically fragile area—Alappuzha District, Southern Kerala India. Nat Resour Res. https://doi.org/10.1007/s11053-018-9419-y

PWD: About Us: Public Works Department, Uttarakhand, India, https://pwd.uk.gov.in/pages/display/2-about-us
Sajikumar N, Remya RS (2015) Impact of land cover and land use change on runoff characteristics. J Environ Manage 161:460–468. https://doi.org/10.1016/j.jenvman.2014.12.041
Satterthwaite D, Huq S, Pelling M, Reid H, Lankao PR (2007) Adapting to climate change in urban areas: the possibilities and constraints in low and middle income nations
Sharma S, Mahajan AK (2018) Comparative evaluation of GIS-based landslide susceptibility mapping using statistical and heuristic approach for Dharamshala region of Kangra Valley India. Geoenvironmental Disasters 5:1–16. https://doi.org/10.1186/s40677-018-0097-1
Shukla S, Gedam S (2019) Evaluating hydrological responses to urbanization in a tropical river basin: a water resources management perspective. Nat Resour Res 28:327–347. https://doi.org/10.1007/s11053-018-9390-7
Singh L, Khare D (2021) Estimation of surface runoff using SCS curve number method coupled with GIS: a case study of Vadodara city 205–216. https://doi.org/10.1007/978-3-030-58051-3_14
Sur U, Singh P, Rai PK, Thakur JK (2021) Landslide probability mapping by considering fuzzy numerical risk factor (FNRF) and landscape change for road corridor of Uttarakhand India. Environ Dev Sustain 23:13526–13554. https://doi.org/10.1007/s10668-021-01226-1
Tian H, Banger K, Bo T, Dadhwal VK (2014) History of land use in India during 1880–2010: large-scale land transformations reconstructed from satellite data and historical archives. Glob Planet Change 121:78–88. https://doi.org/10.1016/j.gloplacha.2014.07.005
Tran TH, Dam ND, Jalal FE, Al-Ansari N, Ho LS, Phong TV, Iqbal M, Le HV, Nguyen HBT, Prakash I, Pham BT (2021) GIS-based soft computing models for landslide susceptibility mapping: a case study of Pithoragarh district, Uttarakhand State, India. Math Probl Eng. https://doi.org/10.1155/2021/9914650
Zeiger SJ, Hubbart JA (2016) A SWAT model validation of nested-scale contemporaneous stream flow, suspended sediment and nutrients from a multiple-land-use watershed of the central USA. Sci Total Environ 572:232–243. https://doi.org/10.1016/j.scitotenv.2016.07.178

Rational Evaluation of Synthetic Fiber Reinforced Frozen Soil

Amit Kumar and Ashwani Jain

1 Introduction

Transport facilities exhibit the prosperity of any nation in terms of development. More convenient the transport will be, the faster the development will be. But, sometimes geological formations of any landscape become bottlenecks and rise the hurdles. These hurdles can be checked by using engineered solutions. Another big challenge in the way of laying pavements is weather condition as local weather can adversely affect the prelaid and newly constructed pavements. These unwanted natural conditions could be precipitation, landslides, or even worse e.g., cyclic freeze–thaw (FT). Why cyclic FT is a worse condition than others? Because precipitation or landslides may not occur frequently, but FT cycles are based on daily weather conditions and may cause significant losses to the prelaid and newly constructed pavements. From the beginning of soil stabilization cement and lime (Dempsey and Thompson 1968; Little et al. 1999) are being used in a common practice to stabilize the fine-grained soils. In mid-80s Walker (Walker 1995) tested over 1500 cement-treated clay blocks and found cement content improved strength and durability characteristics while clay content impaired some engineering properties. Saline water immersed cement treated soil remained strong for 90 days but at 270 days the deterioration had begun. This process surprisingly continued till 6 months in spite the strength gain was got slow down after 28 days (Shihata and Baghdadi 2001). Lime stabilized clayey soil showed a better response against wet-dry (WD) and FT with a less fall in UCS after WD than FT, while basanite and furnace cement treated soft clay showed a greater fall in UCS at earlier stage of WD than later. Coal ash treated expansive soil also showed

A. Kumar (✉) · A. Jain
National Institute of Technology Kurukshetra, Kurukshetra, India
e-mail: Kauleamit0089@gmail.com

A. Jain
e-mail: ashwani.jain@nitkkr.ac.in

K. K. Singh and C. S. Prasad Ojha (eds.), *Sustainable Management of Land, Water and Pollution of Built-up Area*, Society of Earth Scientists Series,
https://doi.org/10.1007/978-3-031-56176-4_3

the same response (Al-Kiki et al. 2011; Kamei et al. 2013; Estabragh et al. 2013). Shorter curing period e.g., 3–7 days may produce good results than longer one e.g., 28 days. Also, it will be useful to determine unsoaked strength prior to any durability test for proper calculation of design service life (Kampala et al. 2014; Ahmed and Issab 2014). Not only cement stabilized laterite soil produced better results against weathering conditions, but also melamine intrusion produced significant effects; as the blend may be used as sub-base or base material after little alteration in proportions (Donrak et al. 2018; Biswal et al. 2019). Although cement, lime and some wastes have appreciable results, but exploitation of natural resources limits their usage as conventional soil stabilizers and seeks new alternatives. Lime enriched waste i.e., eggshell powder (ESP) now has been emerging as a new engineered material and producing considerable results in soil stabilization. This tremendous material is not only being used in soil stabilization but also in concrete production and having good results (Ing and Choo 2014; Yerramala 2014; Tan et al. 2016). Eggshell ash (ESA) may minimize the setting time of cement-based products (Maduabuchi and Obikara 2018). ESA produced better results over red mud in lime stabilized soil while alone ESP replacement by 25% lime improved the strength properties (James and Pandian 2016; Anoop et al. 2017). Micrographs showed 10% ESP produced enough dense material to withstand required UCS with low plasticity (Surjandari et al. 2018). ESP improved the strength properties of clayey and sandy soils (Dayakar 2019). Other natural powdery materials like corn cob ash, snail shell powder, palm kernel shell powder, and sawdust ash have also been proved as low budget stabilization materials to produce low-cost construction components with adequate engineering properties to be used in the developing countries (Siva Sankar et al. 2019; Ayodele and Popoola 2019; Ayodele et al. 2019; Singh and Saini 2022; Singh et al. 2020). Plastic waste amalgam with ESP and lime has also proved the efficacy in improving engineering properties of soil (Pauzi et al. 2019; Alzaidy 2019). Although an extensive literature has been discussed here but till now a little information is available on ESP usage against FT cycles, which is the novelty of this research work.

1.1 Scope

In the present study, waste eggshells were used as natural lime source and dust palliative material i.e., NaCl was also mixed to treat the soil. Polypropylene fibers were used to check the crack propagation after repetitive FT cycles. The test results may be used effectively on site. Since, natural materials have been used in the study, so the research promises an economic and ecological solution over conventional cement and lime stabilization.

Table 1 Index properties of soil

Property	Clay + Silt (%)	G	LL (%)	PL (%)	PI (%)	ISCS	OMC* (%)	MDD* g/cc	OMC# (%)	MDD# g/cc
Value	100	2.6	34	29	05	ML	19.33	1.72	14.40	1.84

*Obtained by Standard proctor test and # Obtained by Modified proctor test

2 Materials and Methods

2.1 Materials

To make research work realistic and feasible; the soil was excavated and procured from the Himalayan region of India, located in Jammu and Kashmir state. The Himalayan province is extremely cold region of India. Primarily Himalayan soil contains stone pebbles, vegetation roots, etc. in its origin form. So, to clean and make the soil suitable for laboratory purposes picking and sieving processes were performed first. The subjected soil was found of ML type as per the plasticity chart proposed by Indian Standard Classification (ISCS) System and other relevant properties of soil have been tabulated in Table 1. The granulometric curve of the subjected soil has been given in Fig. 1. Eggshells were used in powdery form in this study and the typical chemical compositions of ESP proposed by Adogla et al. (2016) have been given in Table 2. 425μ sieved ESP was used to maintain a uniform distribution and homogeneity in matrix. Local vendor-provided product properties of alkali activator (NaCl) have been tabulated in Table 3. To strengthen the soil, polypropylene fibers were procured from the leading and giant Indian company, Reliance Industries Limited, and used in arbitrarily distributed manner. Product specifications of fibers have been given in Table 4.

2.2 Design of Experiments

Additives and their dosage percentages have been given in Table 5. Orthogonally oriented experimental design has been given in Table 6. Taguchi technique has been coupled with another optimization tools, produced relatable results in recent years (Naeemi and Yazdani 2012; Ikeagwuani et al. 2021; Dounane and Trouzine 2020; Gobena et al. 2021; Manni et al. 2021; Benayoun et al. 2021).

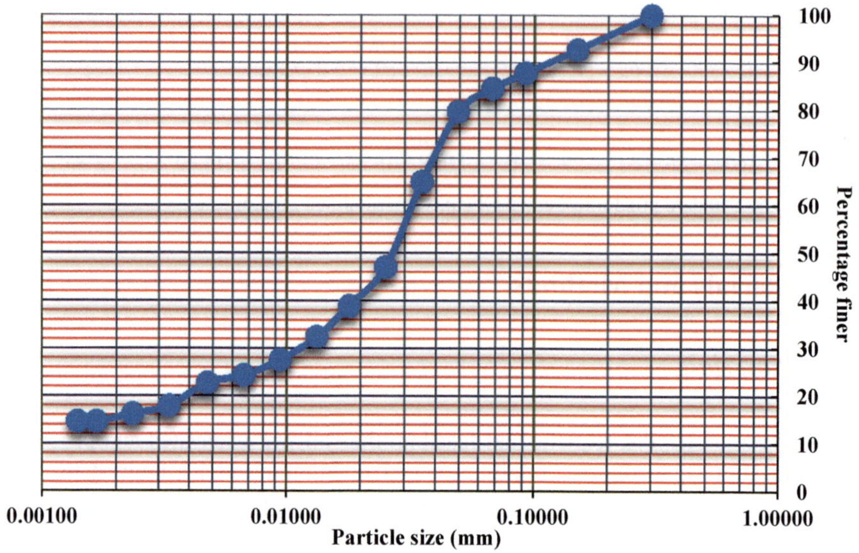

Fig. 1 Granulometric curve of soil

Table 2 Chemical composition of eggshell powder (Adogla et al. 2016)

Group name	Compound	Calcium oxide	Chloride	Aluminium oxide	Silicon dioxide	Manganese oxide	Copper oxide
Chicken eggshell powder	Value (%)	99.83	0.009	0.001	0.001	0.001	0.001

Table 3 Properties of alkali activator

Alkali type	Chemical formula	Color	Physical form	Density at 0 °C (g/cc)	Melting point (°C)	Boiling point (°C)
Sodium chloride	NaCl	White	Crystalline	2.17	801	1465

Table 4 Product specifications of polypropylene fiber

Appearance	Color	Shape	Length (mm)	Relative density (g/cc)	Melting point (°C)	Flash point (°C)
Short length fiber	Pure white	Trilobal	12	0.89–0.94	162–167	> 329

Table 5 Additives and their dosages

Level	Eggshell powder (%)	Polypropylene fiber (%)	Sodium chloride (%)
A	3	0.05	2
B	6	0.10	4
C	9	0.15	6

Table 6 Orthogonally oriented experimental programme

Mix design	Eggshell powder (%)	Polypropylene fiber (%)	Sodium chloride (%)
R0	0	0	0
R1	3	0.05	2
R2	3	0.10	4
R3	3	0.15	6
R4	6	0.05	4
R5	6	0.10	6
R6	6	0.15	2
R7	9	0.05	6
R8	9	0.10	2
R9	9	0.15	4

2.3 *Test Procedure*

To demonstrate the stepwise flow of experimental maneuver, a flow chart has been shown in Fig. 2.

Initially, soil was tested for index properties following various lab procedures specified in the Indian Standards of soil testing. Samples of size (38 mm × 76 mm) were prepared for the UCS test and kept for 21 days for maturing in lab conditions while for durability tests i.e., (Freeze–thaw tests), with the help of a mould (95 mm × 125 mm) and plunger, proctor size fresh samples were prepared and tested after 21 days. Due to long time period consumed in durability testing, many agencies use 7 days UCS test results instead of durability test results, for designing the pavements (Zhang and Tao 2008; Sharma and Dewangan 2019). During whole testing – 25 °C

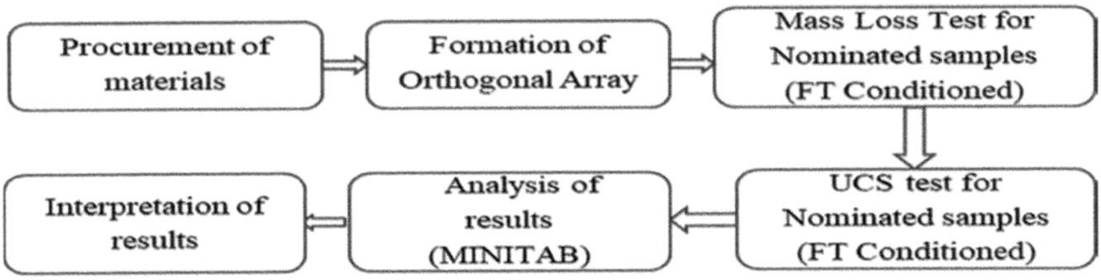

Fig. 2 Flow chart of experimental maneuver

and 20 °C temperature were maintained in freeze–thaw cabinet for the respective freezing and thawing of soil samples. The time gap was maintained about 12 h and 24 h per complete cycle of FT for UCS and durability tests, respectively. Because, after 12 h UCS samples become stable and do not show any significant changes. IS:2720 (Part X) 1991 and IS:4332 (Part IV) 1968 were followed for the testing procedures and calculation purposes of UCS and durability tests, respectively (IS 1968, 1991).

3 Results and Discussion

The present study depicts the effects of freeze and thaw on fine-grained soil being treated with an ameliorating matrix of eggshell powder, sodium chloride, and polypropylene fibers. Samples were kept under 3, 5, and 10 freeze and thaw cycles. Laboratory results-oriented correlations between UCS and Mass loss values for 3, 5, and 10 cycles have been shown in Figs. 3, 4, and 5, respectively.

Analyses of Figs. 3, 4, and 5 showed that there is no any direct correlation between UCS and mass loss of subjected soil under freeze and thaw conditions. The probable reason behind this behavior is the dissimilarity in various parameters e.g., sampling (preparation methods, sizes), testing procedures, and units of measurement of soil properties.

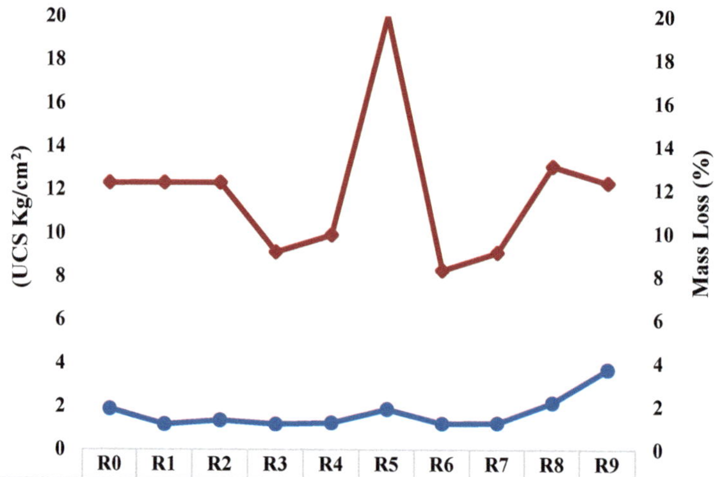

	R0	R1	R2	R3	R4	R5	R6	R7	R8	R9
UCS	1.902	1.187	1.367	1.19	1.251	1.881	1.217	1.236	2.177	3.707
ML	12.3	12.31	12.31	9.12	9.91	20.02	8.28	9.1	13.07	12.3

Fig. 3 UCS versus Mass loss (3 cycles)

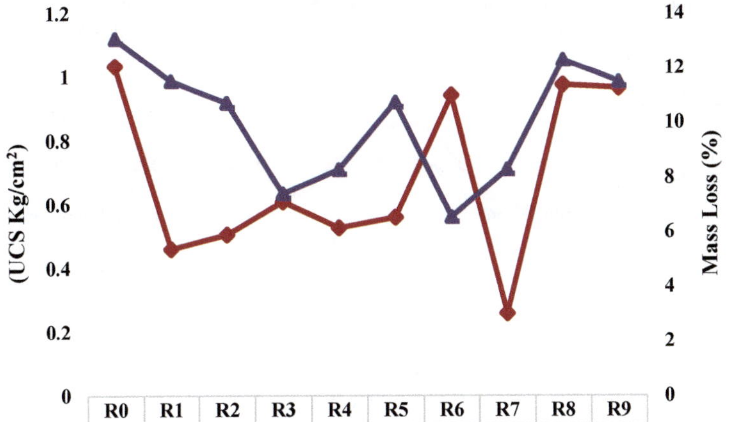

	R0	R1	R2	R3	R4	R5	R6	R7	R8	R9
UCS	1.033	0.46	0.505	0.608	0.526	0.559	0.941	0.258	0.974	0.965
ML	13.07	11.51	10.71	7.4	8.27	10.73	6.53	8.28	12.28	11.49

Fig. 4 UCS versus Mass loss (5 cycles)

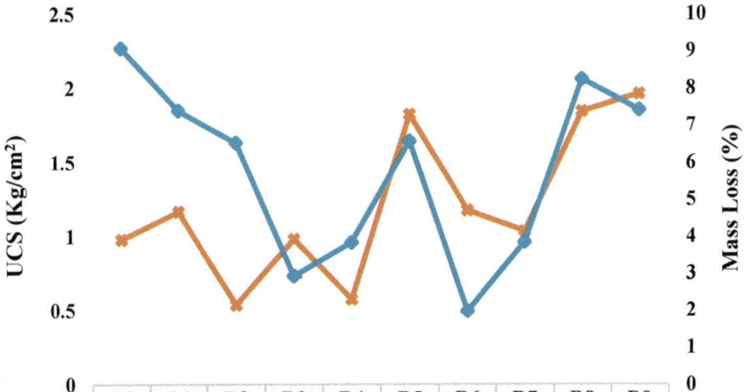

	R0	R1	R2	R3	R4	R5	R6	R7	R8	R9
UCS	0.982	1.167	0.541	0.982	0.578	1.823	1.175	1.036	1.843	1.963
ML	9.08	7.4	6.53	2.94	3.84	6.57	1.98	3.85	8.24	7.41

Fig. 5 UCS versus Mass loss (10 cycles)

Figure 6 showed a cumulative graphical presentation of resistance against mass loss after 3, 5, and 10 freeze and thaw cycles, where, resistance represents the difference between mass loss of a soil sample after repetitive FT cycles from a constant value of about 100%. It can also be concluded that amalgam of the materials rebound the strength and durability properties of the fine-grained soil and made the soil-additive matrix more durable against weathering effects. It can also be observed that material behaved better after every ascending freeze and thaw cycle. This behavior of soil proved the classic theory that believe in the formation of cementitious products

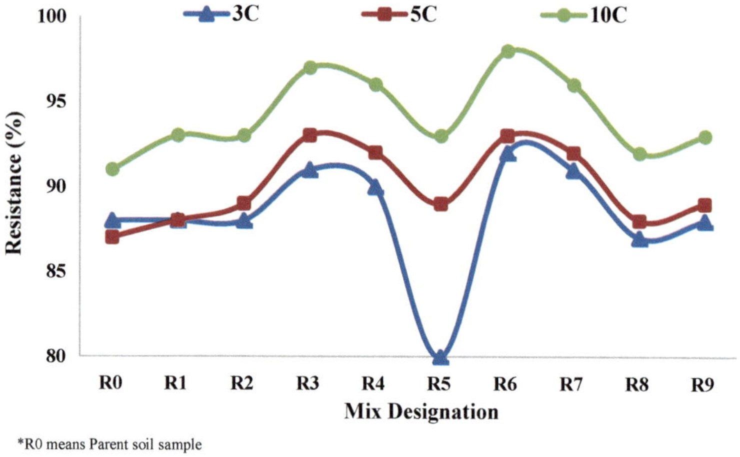

Fig. 6 Durability graph

after the pozzolanic reactions between calcareous materials and water present in the soil (Singh et al. 2022, 2021).

4 Conclusions

The present study interpreted an attempt to establish a correlation between strength and durability properties of fine-grained soil under weathering conditions. Primary conclusions of the above study can be summarized as:

- Treated soil samples performed better against weathering conditions while strength parameters were depleted continuously.
- No direct correlation could be established using different parameters of measurement.
- Available water promoted the pozzolanic reactions and helped to resist the weathering effects.
- Fine-grained soils can be well treated using optimal dosages of selective additives.
- Disposal of waste materials can be reduced by using them wisely as crude resources of engineered materials.

Limitations

- Although the present study has some rational results, still the results may vary in the field due to the differences in procedures of testing at site and laboratory.
- The results are compatible with the literature but may vary under the environment, equipment, and expertise of the performer.

- The present results may not be taken as standard but might be used as a reference for future studies.

Acknowledgements The authors wish to thanks and acknowledge the Council of Scientific and Industrial Research (CSIR), New Delhi for the financial support and Dr. Hari Singh (Professor, Department of Mechanical Engineering, NIT Kurukshetra) for his technical assistance.

References

Adogla F, Yalley PPF, Arkoh M (2016) Improving compressed laterite bricks using powdered eggshells. Int J Eng Sci 5(4):65–70

Ahmed A, Issab UH (2014) Stability of soft clay soil stabilised with recycled gypsum in a wet environment. Soils Found 54(3):405–416. https://doi.org/10.1016/j.sandf.2014.04.009

Al-Kiki IM, Al-Atalla MA, Al-Zubaydi AH (2011) Long term strength and durability of clayey soil stabilized with lime. Eng Technol J 29(4):725–735

Alzaidy MNJ (2019) Experimental study for stabilizing clayey soil with eggshell powder and plastic wastes. In: 2nd International conference on sustainable engineering techniques, IOP Conf Ser Mater Sci Eng 518. 1088/1757-899X/518/2/022008

Anoop SP, Hizana B, Jwoleena PJ, Midhula J, Tharis Muhammed TN, Prasanth S (2017) Potential of eggshell powder as replacement of lime in soil stabilization. Int J Adv Eng Res Sci 4(8):86–88. https://doi.org/10.22161/ijaers.4.8.15

Ayodele FO, Popoola OO (2019) Potential of snail shell and palm kernel shell powders in improving engineering properties of clay. J Appl Sci Environ Manag 23(8):1437–1444. https://doi.org/10.4314/jasem.v23i8.5

Ayodele AL, Oketope OM, Olatunde OS (2019) Effect of sawdust ash and eggshell ash on selected engineering properties of lateralized bricks for low-cost housing. Niger J Technol 38(2):278–282. https://doi.org/10.4314/njt.v38i2.1

Benayoun F, Boumezerane D, Bekkouche SR (2021) Techniques for optimizing parameters of soil nailed vertical cut article in selected scientific papers. J Civ Eng 16(1):131–145. https://doi.org/10.1515/sspjce-2021-0010

Biswal DR, Sahoo UC, Dash SR (2019) Durability and shrinkage studies of cement stabilised granular lateritic soils. Int J Pavement Eng 20(12):1451–1462. https://doi.org/10.1080/10298436.2018.1433830

Dayakar P (2019) Study on strength properties of sand by bio cementation with eggshell. Int J Civ Eng Technol 10(1):2770–2785. http://www.iaeme.com/IJCIET/issues.asp?JType=IJCIET&VType=10&IType=01

Dempsey BJ, Thompson MR (1968) Durability properties of lime-soil mixtures. Highway Research Record 235, Highway Research Board, Washington D.C., pp 61–75

Donrak J, Horpibulsuk S, Arulrajah A, Kou H, Chinkulkijniwat A, Hoy M (2018) Wetting-drying cycles durability of cement stabilised marginal lateritic soil/melamine debris blends for pavement applications. Road Mater Pavement Design 1–10. https://doi.org/10.1080/14680629.2018.1506816

Dounane N, Trouzine H (2020) Integrated approach of ANN and Taguchi algorithms for soils swelling estimation. Int J Sustain Build Technol Urban Develop 11(4):244–257. https://doi.org/10.22712/susb.20200019

Estabragh AR, Pereshkafti MRS, Parsaei B, Javadi AA (2013) Stabilised expansive soil behavior during wetting and drying. Int J Pavement Eng 14(4):418–427. https://doi.org/10.1080/10298436.2012.746688

Gobena JA, Kumar G, Suppiah S (2021) Optimization of Atterberg limits of treated expansive soils with Taguchi method. Mater Today Proc 43(2):2127–2133. https://doi.org/10.1016/j.matpr.2020.11.894

Ikeagwuani CC, Agunwamba JC, Nwankwo CM, Eneh M (2021) Additives optimization for expansive soil subgrade modification based on Taguchi grey relational analysis. Int J Pavement Res Technol 14:138–152. https://doi.org/10.1007/s42947-020-1119-4

Ing DS, Choo CS (2014) Eggshell powder: potential filler in concrete. In: 8th MUCET 2014, Malaysia

IS: 4332-Part IV: Specifications for wetting and drying, and freezing and thawing tests for compacted soil cement mixture. Bureau of Indian Standards, New Delhi, India (1968)

IS: 2720-Part 10: Indian standard methods of test for soils: determination of unconfined compressive strength. Bureau of Indian Standards, New Delhi, India (1991)

James J, Pandian PK (2016) Development of early strength of lime stabilized expansive soil: effect of red mud and eggshell ash. Acta Technica Corviniensis Bulletin of Engineering Tome IX, Fascicule (1):93–100

Kamei T, Ahmed A, Ugai K (2013) Durability of soft clay soil stabilized with recycled Bassanite and furnace cement mixtures. Soils Found 53(1):155–165. https://doi.org/10.1016/j.sandf.2012.12.011

Kampala A, Horpibulsuk S, Prongmanee N, Chinkulkijniwat A (2014) Influence of wet-dry cycles on compressive strength of calcium carbide residue–fly ash stabilized clay. J Mater Civ Eng 26(4):633–643. https://doi.org/10.1061/(ASCE)MT.1943-5533.0000853

Little D, Males E, Prusinski J, Stewart B (1999) Cementitious stabilization. Millennium paper, Transportation Research Board, Washington D.C.

Maduabuchi MN, Obikara FO (2018) Effect of eggshell powder (ESP) on the strength properties of cement-stabilization on Olokoro Lateritic soil. J Waste Manage Xenobiotics 1(1):1–5. https://doi.org/10.23880/oajwx-16000103

Manni A, Saviano G, Bonelli MG (2021) Optimization of the ANNs predictive capability using the Taguchi approach: a case study. Mathematics 9(766):1–16. https://doi.org/10.3390/math9070766

Naeemi A, Yazdani M (2012) Application of Taguchi method in optimization of soil-nailed excavation design. Int Soc Soil Mech Geotech Eng, ANZ Conf Proc 1389–1393

Pauzi NIM, Aziz MAAMA, Ismail MS, Omar H (2019) Waste soil improvement using eggshell powder and lime at open dumping area. Int J Civ Eng Technol 10(4):268–277. http://www.iaeme.com/IJCIET/issues.asp?JType=IJCIET&VType=10&IType=4

Sharma A, Dewangan N (2019) A review paper on strength and durability properties of stabilised soils. Int J Res Appl Sci Eng Technol 7(VIII):739–752

Shihata SA, Baghdadi ZA (2001) Long-term strength and durability of soil-cement. J Mater Civ Eng 13(3):161–165

Singh MJ, Feng W, Xu DS, Dubey M, Borana L (2022) Long-term elastoviscoplastic behavior of fly ash-blended Indian montmorillonite clay in oedometer conditions. Int J Geomech 22(3):1–14. https://doi.org/10.1061/(ASCE)GM.1943-5622.0002290

Singh G, Saini B (2022) Nanomaterial in cement industry: a brief review. Innov Infrastructructure Solutions 7(45). https://doi.org/10.1007/s41062-021-00649-z

Singh G, Singh M, Saini B (2021) Role of FRP in developing sustainable infrastructure-a review. In: Ashish DK, de Brito J, Sharma SK (eds) 3rd International conference on innovative technologies for clean and sustainable development. ITCSD 2020. RILEM Bookseries, vol 29. Springer, Cham. https://doi.org/10.1007/978-3-030-51485-3_16

Singh MJ, Borana L, Feng W, Xu DS (2021) Long-term swelling characteristics of montmorillonite clay with and without fly ash: wetting-drying cycle influence in 1D oedometer condition 50(6). https://doi.org/10.1520/JTE20210248

Siva Sankar V, Arumai Raj PD, Janaki Raman S (2019) Stabilization of expansive soil by using agricultural waste. Int J Eng Adv Technol 8(3S):154–157

Surjandari S, Dananjaya RH, Jauharotus SE (2018) The effect of eggshell powder on the compression strength of fine-grained soil. MATEC Web of Conferences, ICRMCE 2018 195(03011):1–6. https://doi.org/10.1051/matecconf/201819503011

Tan YY, Doh SI, Chin SC (2016) Durability analysis of natural lime concrete. In: Yacob NA et al (eds) Regional conference on science, technology and social sciences. https://doi.org/10.1007/978-981-13-0074-5_26

Walker PJ (1995) Strength, durability and shrinkage characteristics of cement stabilised soil blocks. Cement Concr Compos 17:301–310

Yerramala A (2014) Properties of concrete with eggshell powder as cement replacement. The Indian Concr J 94–102

Zhang Z, Tao M (2008) Durability of cement stabilized low plasticity soils. J Geotech Geoenviron Eng 134(2):203–213. https://doi.org/10.1061/_ASCE_1090-0241_2008_134:2_203_

Geological Characterization of Bhama Askhed Dam Site

D. S. Aswar, P. B. Ullagaddi, and S. D. Ambadkar

1 Introduction

The safety, stability and effectiveness of a dam structure, depend largely on the geological conditions of the foundation. The specifications of geologic/geotechnical requirements for expected foundation performance should be ensured. For the potential dam sites, competent bedrock at reasonable depth should be available. The bedrock should also be free from seams, joints or fractures to avoid an objectionable amount of leakage. The decisions, such as depth of foundation, need and the extent of treatment for weaker zones should be assessed rationally based on the interpretation of investigation data. Its successful implementation will result in, proper project decisions for significant improvements in design, and construction, resulting in the durability of the project.

The conventional visualization of geological structures and corresponding geoparameters through geological cross-sections, enhance the understanding of the geological fluctuations and the spatial variation of geological parameters. An engineering geological assessment of the dam area has to be carried out to check the project's feasibility (Berhane et al. 2013). The results of borehole data, Lugeon test, results of downhole geophysical methods and hydrogeological investigation can be used in the development of the dam's preliminary design and confirmation of the foundation suitability for the proposed dam structure (Marley et al. 2007). Geological and geophysical investigations can be conducted in the abutment and reservoir areas to

D. S. Aswar (✉)
Sinhgad College of Engineering, Pune 411041, India
e-mail: dineshaswar@gmail.com

P. B. Ullagaddi
S.G.G.S. Institute of Engineering and Technology, Nanded 432606, India

S. D. Ambadkar
G H Raisoni University, Amravati 444701, India

K. K. Singh and C. S. Prasad Ojha (eds.), *Sustainable Management of Land, Water and Pollution of Built-up Area*, Society of Earth Scientists Series,
https://doi.org/10.1007/978-3-031-56176-4_4

evaluate the causes of leakages. Site exploration with boreholes and packer tests along with rock mass discontinuity analysis can be used for assessing the causes of leakages (Berhane 2010). Field studies cover discontinuity surveying and sampling for rock mass strength determination. The core logging data could be used to construct geologic cross-sections and subsurface profiles. It provides the design engineers with critical data early in the project and also helps to identify any additional data needs (Rogers 2012).

2 Methodology

The paper presents the site characterisation for the Bhama Askhed irrigation project. The energy dissipating assembly (EDA) failure at the site emphasizes, the deficiency in exploration data integration, interpretation and appraisal of required foundation treatment. Also, the excavated tail channel at the project site is experiencing rapid erosion along the alignment.

The study consists of, geological site characterization, and evaluation for the project decision and future prognosis. The geological, and geotechnical data from various phases and sources of investigation is integrated, reanalyzed and interpreted for geological prediction of the area. The intent is to characterize the dam project site within the context of its surrounding geology and geotechnical characteristics. The study consists of distinguishing critical locations along the dam alignment and tail channel, characterized by specific geological or geotechnical scenarios related to rock mass parameters.

All the available core log data was critically reanalyzed for characterization of the foundation stratum. The depth-wise lithology, core recovery, RQD, jointing and percolation data was compiled and reanalyzed, reinterpreted and presented in a more concluding format as per the International Society for Rock Mechanics (ISRM). The tail channel erosion studies are conducted and accordingly, recommendations are given.

2.1 Study Area Bhama Askhed Irrigation Project

The project, is on the river Bhama, at village Waki (18°15′N,73°43′E), Pune, Maharashtra (India). It's a 51.125 m high and 1425 m long earthen dam with an FRL (full reservoir level) storage capacity of 230.47 million cubic meters and a 55.5 m long ogee side spillway. The details of geological exploration stages and drilled holes are shown in Table 1.

Detailed core logging was carried out for the preparation of Lithologs. Based on the site investigation data, foundation and grouting levels were confirmed. The core samples data reveal the complex nature of the substratum at the exploration site,

Table 1 Details of drill holes (DH)

Exploration stages	No. of bore holes	Depth (m)
Bhama Askhed (BA) (waki stage-I)	30	611.51
BA (stage II)	24	634.015
COT/ICPO/APPL (stage III)	11	198.79
Total along tail channel (TL)	23	304.485
Total drill holes (DH)	88	1748.80

Abbreviations Cut-off Trench (COT), Irrigation cum Power Outlet (ICPO); Approach Channel (APPL)

emphasizing the variation in lithology, layers thickness, fracture intensity and the engineering properties of the rocks.

Lithology at Bhama Askhed. The major rock type in the study area is compact basalt and amygdaloidal basalts (A.B.) (Kulkarni 1994; Kulkarni et al. 2007). The compact basalts have superior mechanical properties to amygdaloidal basalts however, the presence of the divisional planes demands careful consideration (Gupte 1980). Divisional planes such as joints, bedding planes etc. are the source of trouble for the dam foundation. The joints need grouting to improve their strength and to make them watertight. Table 2 gives the percentage distribution of overburden soil (O.B.) and various rock types observed at the dam site.

Among the identified lithologic classes from core logs, few rock types have the potential to cause a nuisance for engineering applications. Hydrothermally altered Tachylytic Basalts, Volcanic Breccia (V.B.) with lava matrix are prone to rapid deterioration on exposure to the atmosphere Fig. 1. Their proportion is 2.91% as explained in Table 2.

Core Recovery. Core recovery indicates the physical condition of the rock mass. Rocks giving high core recovery values are considered good from the engineering point of view. Poor core recovery may be due to various causes, such as poorly consolidated rock, sheared zones, clay-filled seams or weathered material, the inclination of the hole related to the rock foliation and faulty drilling techniques.

Rock Quality Designation (RQD). RQD is a measure of rock mass jointing/ discontinuity levels and also describes the joint spacing. The RQD is a more sensitive and consistent indicator of general rock quality compared to the gross recovery percentage (IS 11315 (Part 12) 1992). RQD is a quality assessment parameter for dam design. RQD is the rock quality index for preliminary assessment. It depends on the strength and number of natural discontinuities of the rock mass (Deere and Deere 1988). The low RQD suggests poor drilling method, improper core handling, stress relief or slaking, jointed, sheared or weathered rock. The average recovery% and RQD% variation for the right abutment, left abutment and the dam middle portion is 3.87%, whereas it is quite high as 63.08% for the approach channel indicating its fractured state of rock. The data analysis (Tables 3 and 4) evidenced that there is no direct correlation between RQD and core recovery values. Classification and the distribution of RQD% values observed at the site are given in Tables 3 and 4

Table 2 Consolidated rock distribution (length in meters)

No.	Location	O.B	C.P.B	A.B	V.B	T.B	Total
1	Waki main dam line M.D.L. (21 DH)	46.55	291.67	92.37		9.01	439.60
2	BA M.D.L. (21 DH)	43.70	500.43	39.20	3.60	3.14	590.07
3	COT and ICPO (6 DH)	4.17	81.53	30.88	14.78	3.38	134.74
4	Waki D/s of MDL (9 DH)	13.60	98.58	58.24		1.49	171.91
5	BA tail channel TL (3 DH)	6.95	35.13	1.87	0.00	0.00	43.95
6	BA-1996 TL (23 DH)	47.90	227.90	21.48		7.09	304.37
7	APPL (5 DH)	0	57.17	3.12	1	2.76	64.05
Total		162.87	1292.41	247.16	19.38	26.87	1748.7
% Thickness		9.31	73.90	14.13	1.11	1.54	100
% Rock Drilled		0	81.50	15.59	1.22	1.69	100

Abbreviations Main dam line (M.D.L); Drill Hole (DH), Bhama Askhed (BA); Downstream (D/s)

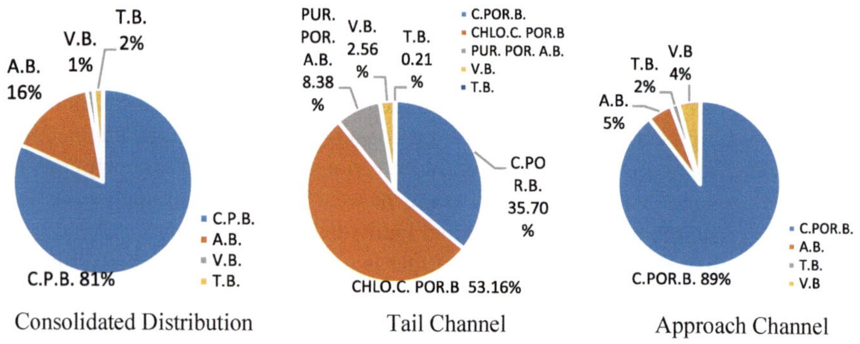

Fig. 1 Percentage distribution of rock types at the project site. *Abbreviations* Compact Porphyritic Basalt (C.P.B/C.POR.B), Amygdaloidal Basalt (A.B.), Volcanic Breccia (V.B.), Tachylytic Basalt (T.B.), Chlorophaeitic Compact Porphyritic Basalt (CHLO.C.POR.B.), Purple, porphyritic Amygdaloidal Basalt (PUR.POR.A.B.)

According to RQD classification, the rock mass in the approach channel is classified as of "poor quality". For the approach channel, RQD% of 26.77% indicates fractured/ jointed poor-quality rock there.

Table 3 Core recovery and RQD values across the dam alignment

No.	Location	Exploration stage	Recovery %	Avg. recovery %	RQD %	Avg RQD %	Avg percolation (Lugeon)
1	Left abutment Chainage (CH 840 to CH 340)	Stage-I	–	91.61	93.18	91.51 (Excellent Rock)*	–
		Stage-II	92.63		89.84		1.39
		Stage-III	90.58		–		–
2	Middle portion (CH-340 to CH -145)	Stage-I	–	85.78	85.69	80.45 (Good Rock)*	–
		Stage-IIBA	85.78		75.20		4.11
3	Right abutment (CH-145 to CH -645)	Stage-I	–	89.63	89.93	84.50 (Good Rock)*	–
		Stage-II	89.63		79.06		0.07

*Deere and Deere (1964) RQD% based rock classification
Abbreviation Chainage (CH)

Table 4 Core recovery and RQD values for the tail channel section

No.	Location	Avg. recovery %	Avg. RQD %	Avg. percolation (Lu)
1	Approach channel (stage-IV APPL)	72.51	26.77 (Poor rock)*	9.02 Moderate**
2	Tailrace channel (stage-III TL)	93.03	89.82 (Good rock)*	Low**

*Deere and Deere (1964) rock classification based on RQD% value, **Camilo Quinones Rozo (2010) Lugeon classification

Distributions of Discontinuities with Depth at the Dam Site. The intensity of joints is the number of joints per unit distance perpendicular to the plane of joints in the set. It partly indicates the extent to which discontinuities affect the rock strength properties. Highly jointed rocks will be inherently weaker and more prone to failure than rocks in which joints are widely spaced.

Table 5 shows the number of joints observed in various stages of investigations. The joint frequency (Joint/m) increases significantly between 2 and 15 m. The correlation between joint frequency and depth shows a reduction in joint frequency with increasing depth. After 40 m it decreased significantly. The joint frequency trend lines for the tail channel display very irregular behaviour. However, the overall trend for the tail channel indicates a decrease in joint frequency with depth.

Water-tightness. The stability of the dam may be threatened by uplift pressure and erosion due to seepage. A steady flow through leaky rocks may have an erosive and weathering effect, which will gradually widen the openings responsible for leakage. The Lugeon test evaluates the average hydraulic conductivity for the determination of the critical permeability and grouting requirements (Camilo Quinones Rozo 2010). At Bhama Askhed dam site permeability was evaluated by double packers Lugeon Test (IS 5529 2006). The rock's hydraulic conductivity is affected by the joint spacing

Table 5 Jointing details (natural joints) in different rocks

Sr. No.	Location/No. of drill hole	Joints	C.P.B	A.B.W.I	A.B.G.I	G.P.B	Other
1	M.D.L. (Waki/18)	274	224	2	38	7	3
2	Tail channel (Waki/09)	215	116	96	0	0	3
3	M.D.L. (BA/13)	57	570	4	1	1	0
4	Tail channel (BA/25)	331	331	0	0	0	0
Total joints		1396	1241	102	39	8	7
% joints		100	88.90	7.31	2.79	0.57	0.50

Abbreviations Main dam line (M.D.L), Bhama Askhed (BA), Compact Porphyritic Basalt (C.P.B.), Amygdaloidal Basalt with White infillings (A.B.W.I.), Amygdaloidal Basalt with Green infillings (A.B.G.I.), Giant phenocryst Basalt (G.P.B.)

and infilling characteristics (Goodman 2013). For the permeability tests in hard rocks, the water intake is generally due to joints and fractures. Hence, the permeability corresponds to the joints and fractures. The Lugeon value indicates the condition of rock discontinuities.

The foundation rocks must be watertight or else, appropriate steps should ensure to prevent seepage through them. The objective of grouting is to control the rate of seepage and uplift through the substratum. The lugeon test data is used to demarcate the high permeability zones (permeability > 3 Lu i.e. lugeon) by geomodelling, which needs seepage reduction measures. Table 6. shows the details of lugeon values at the dam site.

The high percolation value for the approach channel (9.02 Lu, few partly open discontinuities), and tailrace channel (3.7 Lu) (Table 4) is quite high for the spillway section. Figure 2 shows high percolation along Approach Channel for DH-APPL-2 To eliminate the possibility of misleading conclusions being drawn from percolation tests the results of each test section are considered in conjunction with the core from the test section and results inconsistent with the nature of the core rocks are discarded. Figure 3. shows a similar correlation of RQD and associated lugeon value carried out representing the consistency of lugeon values. The lugeon values give

Table 6 Lugeon test values for Bhama Askhed project site

Approach channel	Tests	% distribution	Rock mass discontinuities
Impervious (0–3 Lu)	10	30.30	Few partly open
Low permeability (03–10 Lu)	4	12.12	Tight
Medium permeability (10–30 Lu)	19	57.58	Some open
High permeability (30–60 Lu)	0	0.00	Some open
Very high permeability (>60 Lu)	0	0.00	Many open
Total number of tests	33 no	100.00	
Permeability > 3 Lugeon	23 no	69.70	
Permeability > 10 Lugeon	19	57.58	

supplementary data, which may be presented as a log, parallel with that for total core recovery, frequency and RQD, etc. The low RQD values suggesting high fracture frequency or close spacing of discontinuities along with the high permeability values at the same depth are indicative of the open nature of fracture apertures. The representation of Lugeon and RQD % Variation with Depth for the deepest borehole in approach channel DH-APPL-2 is an indication of open fractures at the depth of around 25 m which need to be grouted Fig. 3. As observed earlier in Tables 3 and 4, the average percolation value for the left and the right abutment (excluding the spillway section) was 1.39 and 0.07 Lugeon (Lu) respectively. However, values are 4.11, 9.02 and 3.69 Lu respectively for the middle portion, approach channel and tailrace channel, which is higher than the permissible value of 3 Lu. The distribution of lugeon values above 3 Lu (i.e. pervious strata) is lower for both abutments than for the middle portion and spillway section. The right abutments' rock mass is classified as impervious however the spillway section at the right abutments is permeable to medium permeable with the presence of some open discontinuities. On average, the foundation rock mass at the right abutment is classified as moderately permeable (average permeability of 4.26 Lu). There are frequent conditions of no pressure developed during percolation tests due to heavy water loss on numerous occasions for these locations.

The discontinuity surveys, recovery, RQD and permeability tests, have pointed to the requirement of substratum treatment (Uromeihy and Farrokhi 2011). The bulk of the Lugeon values (69.7%, Table 6) in the approach section and tailrace channel (at EDA) between tail channel CH −50 m to +112 m across the spillway were higher than 10 Lugeon (Fig. 2) indicating the possibility of excessive foundation leakages. The high permeability of the middle portion and spillway sections is due to the presence of fractured or jointed rocks (as shown in recovery/RQD Tables 3 and 4). This indicates grouting treatment requirements for these sections especially

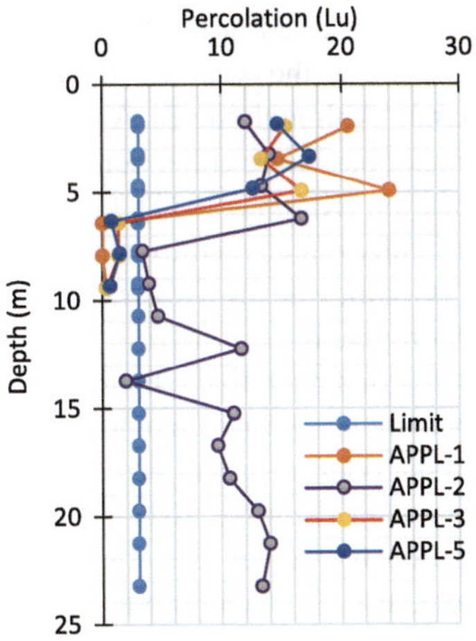

Fig. 2 Percolation along approach channel

Fig. 3 Lugeon, RQD% variation (APPL-2)

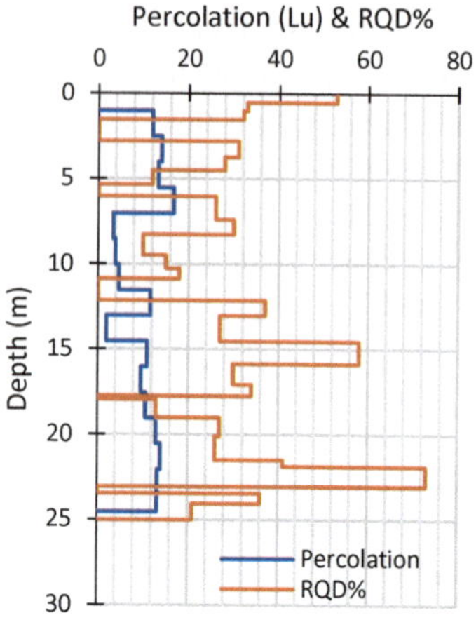

for the spillway section, to improve the rock mass units preventing leakage through the spillway foundation.

Normally, the permeability decreases with depth. However, Fig. 2. shows inconsistent permeability relationship with depth. The drilling data analysis showed that, as there are few scattered values. There was no straightforward relation between RQD and permeability values. The lugeon values lower than 0.1 Lu around 40 m depth for the spillway section. Under such diverse conditions, the occurrences of sound foundation grade bedrock, weathered rock, fracture zones and percolation values should be considered together for checking grouting requirements. As per IS:6066 (IS 6066 1994), for dams higher than 30 m, curtain grouting is required to reduce seepage for foundation rock mass with percolation greater than 1 Lu. For the Rock below the cutoff trench, the limiting values of permeability are between 1 and 3 Lu.

Seepage Problem at Spillway Section. Continuous high seepage of 10–15 L/s was observed at the EDA failure portion (RD 83 m.). At RD 83 m, there was a junction of the weathered, hydrothermally altered jointed volcanic breccia and the basalt. Although the EDA concrete panels were firmly anchored to the bedrock, the anchor bars were either bent or broken and had been partially withdrawn at the failure location (Kamble 2011). Given this failure, the geology at the spillway section was critically reanalyzed to investigate the failure.

Geology at the Spillway Section. The dam is with a side spillway, situated on the chlorophacitic porphyritic basalt bedrock (Kamble 2011). The tailrace channel flow is diverted back to the main river course, with the artificially excavated waterway. A dyke in the tailrace channel of 2–2.5 m width crosses the spillway. The dyke portion on the dam upstream has been grouted whereas the downstream portion remained untreated (Kamble 2011). Geology across the spillway section (Fig. 4) along the approach channel axis consists of porphyritic basalt flows separated by the uneven thickness of volcanic breccia. The borehole cross-section along with the spillway shows the existence of soft brown hydrothermally altered porphyritic amygdaloidal basalt 35 m depth (DH-111). In general, the correlation between core recovery,

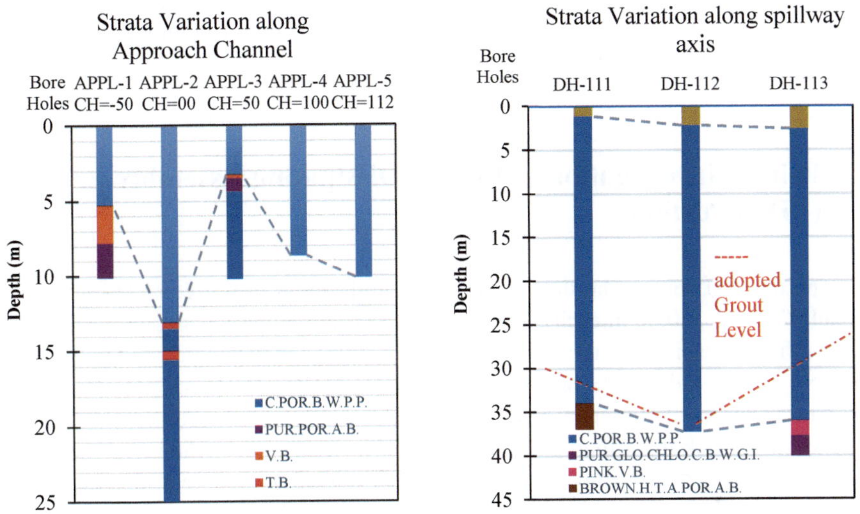

Fig. 4 Borehole cross-section along approach channel and spillway section. *Abbreviations* Compact Porphyritic Basalt with Plagioclase phenocrysts (C.POR.B.W.P.P), Purple porphyritic Amygdaloidal Basalt (PUR.POR.A.B), Volcanic Breccia (V.B.), Tachylytic Basalt (T.B.), Purple Glomeroporphyritic Chlorophaeitic Compact Basalt with Green infillings (PUR.GLO.CHLO.C.B.W.G.I.), Pink Volcanic Breccia (P.V.B.), Brown Hydrothermally Altered Porphyritic Amygdaloidal Basalt (BROWN. H.T.A.POR.A.B.)

and RQD trends with rock depth shows an increase in these values with depth and high values prevails at greater depths. Thus, indicating suitable geological strata for engineering applications. However, for the approach channel, there are significant inconsistencies of RQD values even at greater depths indicating the fractured and jointed state of rocks. For the approach channel, both recovery and RQD values are inconsistent and decreasing with depth signalling jointed and fractured formation below.

For the spillway section, the average values for both Recovery and RQD are around 80% however, the deepest borehole DH-113 shows inconsistency at depth. Spillway Cross-section also shows, inconsistent geology with, Soft H.T.A. Porphyritic Amygdaloidal Basalt below 35 m depth (DH-111), Pink V.B. pocket at a depth around 37 m (DH-113). The data analysis for the spillway section indicates, the need for foundation grouting up to the depth of 40 m, which is greater than the adopted/ executed average depth of 35 (20–25) m. For the later reach of the tail channel, however, the recovery and RQD values show a certain improvement.

Figure 2.3 show the percolation variation with the depth along the approach channel.

The average permeability for the approach channel and the tail channel was above 9 Lu and about 3.67 Lu respectively. For the tail channel, there were several sections with no test pressure developed due to heavy water losses. Both approach channel and tail channel sections show open-jointed fractured formations and loose strata.

The entire approach section and tail channel near the spillway (including EDA) need proper treatment to prevent seepage and uplift pressure.

3 Failure Investigation at Energy Dissipating Assembly (EDA) Portion

During the 2005 monsoon, there was a structural failure of thirteen EDA concrete panels ($7 \times 11 \times 0.30$ m each) because of underneath high seepage, uplift pressure and spillway water discharge forces. Kulkarni (2006) conducted detailed geological mapping of the failure section and tailrace channel. The geological investigation of the tailrace channel alignment with 23 boreholes indicates fractured closely jointed and splitting rock mass there. The tail channel is situated over the untreated, fractured Chlorophaeitic Porphyritic Basalt (CHLO.POR.B.), capable of disintegration on exposure. This stratum is over untreated soft, erodible, Hydrothermally Altered Volcanic Breccia (H.T.A.V.B.) with a Red Tachylytic Basalt (R.T.B.). The results of the borehole logging indicated the weak and permeable zone at the spillway section as reinterpreted in Fig. 4. The reanalysis of litho-logs also confirms the presence of red breccia at the depth of 36 to 38 m, which may have caused the seepage at failure section (RD 83 m). It also concluded a need for treatment to the underlined jointed fractured, compact basalt and weathered H.T.A.V.B. with R.T.B. The uplift due to seepage through pervious tailrace channel substratum rocks has resulted in the panel displacement. Thus, the EDA portion has failed because of high seepage under the spillway section due to inadequate foundation treatment and failure to prevent excessive seepage pressure.

To identify the causes of failure and to suggest remedial measures, Central Water and Power Research Station (CWPRS), Pune carried out the investigation. The nuclear logging and tracer tests were conducted to identify the seepage path to the failure portion. Calliper logging was also attempted to check the extent of joints and fractures in the boreholes (Kamble 2011; CWPRS, Pune 2008). The weaker red breccia zone was traced at around 34 m depth by nuclear logging. The tracer studies revealed that the dye injected in the boreholes where the prominent cracks/weakness zones were detected by the nuclear logging arrives at the leakage point downstream of the spillway section. Thus showing the interconnectivity between failure location and weaker red breccia in the spillway foundation. The findings of these investigations compare well with the reinterpretation of the geoparametric analysis carried out in this research work.

3.1 Tail Channel Erosion

The success of flank or sidesaddle spillway depends on the suitability of excavated tail channel rocks to resist erosion by overflowing flood. If not checked the tail channel erosion will gradually progress towards the spillway, endangering spillway safety by undermining foundation rocks. The tail channel erosion does not pose an immediate threat; spillway safety will be endangered with time. At places where geological studies indicate serious erosion of the tail channel, either the proposal for a side spillway needs to be abandoned (Kulkarni 2007). The spillway foundation substratum and excavated tail channel at Bhama Askhed consist of block jointed, hydrothermally altered basalts, volcanic breccias and red tachylytic basalt that may lead to seepages through the spillway foundation and rapid erosion along the tail channel alignment. The profile survey was carried out to verify the Tail channel erosion.

Erosion along the Tail channel centerline shows the deepening due to erosion along with the longitudinal centre line profile Fig. 5. The EDA end and weathered tailrace channel outcrop experiencing gradual erosion. The extent of this erosion was quantified through a comparison of the designed initial tail channel RLs (reduced levels-1997) with the RLs obtained by the conducted profile and cross-section survey (2015) and also with the failure RLs (2006) over the 351 m of the tail channel.

The available project tail channel profile RL data and conducted survey RL data are used to model the surface grid. Volumetric calculations from the grid models (Surfer ™) assess the tail channel erosion. Table 7 depicts the calculated quantity of tail channel erosion with time (Kulkarni 2007) (Aswar and Ullagaddi 2022) (Aswar 2018). Tail Channel material of 12,885.79 m^3 had been eroded due to the tail channel erosion approximately over the span of fifteen years. Figure 6 show the variation of contour/elevation for the tail channel due to erosion. Figure 7 represents a temporal stack of eroded surfaces; these show the erosional features and the deepening of the tail channel.

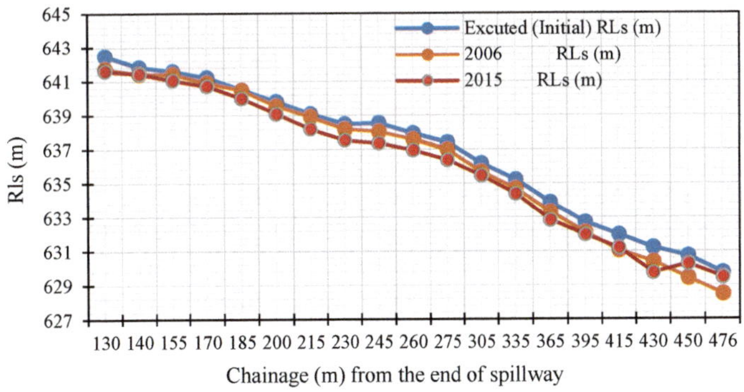

Fig. 5 Erosion along the tail channel centerline over time

Table 7 Tail channel erosion

Years	2000–06	2006–15	2000–2015
Net volume of erosion	9866.44 (m^3)	3340.28 (m^3)	12,885.79 (m^3)

Fig. 6 Contour map for tail channel Erosion (EDA end RL. 125 m to RL. 475 m)

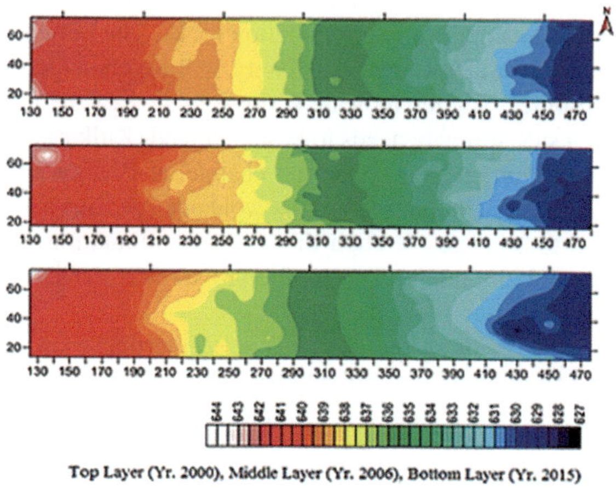

Fig. 7 Stacked 3-D surface model for excavated top and eroded tail channel

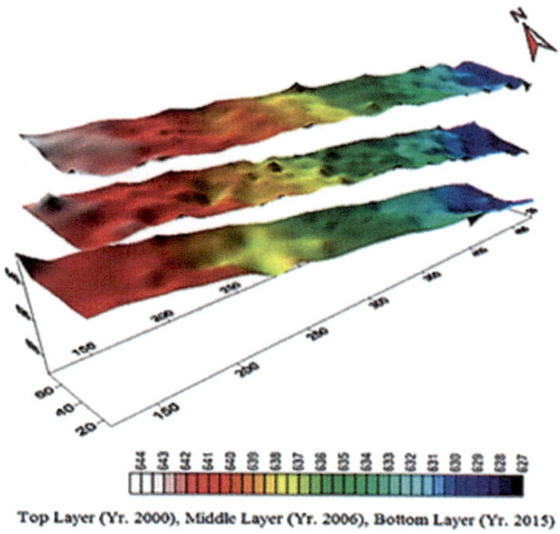

In comparison between the tail channel RL of 1997 and 2015, it is observed that RL is decreasing for every chainage from the end of spillways. Over the temporal scale from 2000 to 2015, there is a substantial increase in erosion as indicated by the backward progression of low elevation values towards the spillway section. It indicates the continued unsafe tail channel erosion. The grid models affirmed the gradual progressive erosion towards the spillway section. The erosion needs to be checked to prevent future potential seepage through exposed Tachylytic formation. Such exposed red tachylytic basalts, therefore, have to be sealed by blanketing to prevent further weathering. Considering tail channel erosion potential further appropriate measures are suggested in the recommendation section. The dolerite dyke across the spillway in the approach and tail channel needed to be sealed, to prevent further seepage, erosion and uplift issues. The performance of EDA relief holes also needs to be ensured, for effective reduction of the uplift pressure.

4 Conclusion

1. The dam site characterization is performed with a detailed analysis of geoparameteric variation and interpretation. The critical location along the dam alignment at the spillway section and also along the tail channel, characterized by specific geological scenarios related to rock mass parameters is identified which also evaluates the causes of failure of the EDA portion. Based on the study of the foundation substratum depth of grouting treatment required at the spillway section, the need for treatment at the tail channel to prevent further erosion is identified.
2. The Energy Dissipating Assembly (EDA) failure emphasizes, the deficiency of exploration data integration, interpretation and appraisal of required substratum treatment.
3. The executed grouting details for the project site were not available, hence the conclusive remarks about the location extent and efficiency of carried out grouting could not be enquired. However, the failure of the EDA portion has emphasized the concerns about the degree and depth of grouting adopted. Leakage could easily have been prevented if the proper grouting to the required depth had been carried out before the reservoir filling.

All the favourable geological conditions are difficult to meet at the available project site. However, judicious project decisions should be made based on the analysis of available investigation data.

5 Recommendations

The interpretation of the percolation and drilling data analysis concludes that the project site is geologically complex and there is a need to address future performance-related issues for long-term use. Following are the recommendations made based on the research findings.

1. The majority of the tail channel section is potentially venerable to erosion and needs systematic reservoir management, using ICPO and a newly constructed jack well considering the erosion conditions at the tail channel. Protective measures are to be adopted for erosion may consist of the construction of a series of fall structures, and water cushioning, for energy dissipation. Also, guide

walls and the bed concrete apron can be used for bed protection. Red tachylytic formations at the end of the EDA and the tailrace channel are witnessing gradual erosion. These are required to be sealed by blanketing to avoid further weathering.
2. The relief holes suggested by CWPRS may not prove to be the long-term solution for relieving the seepage forces. The turbidity of seepage witnessed is directly linked with the fines passing downstream through relief holes. The volcanic breccia with red tachylytic basalt in the formation consists of clay minerals, prone to weathering due to seeping water. The quantity of fines passing downstream is likely to increase due to weathering of weak, permeable, low-density rock zones in the foundation. Thus, choking the porous drains with leaching materials. The reliable performance of filter and drain holes needs to be assured to relieve the uplift pressure.

References

Berhane et al (2013) Geological challenges in constructing the Geba dam site. North Ethiopia Eng Geol Environ 72. https://doi.org/10.1007/s10064-013-0480-9

Marley M et al (2007) The geotectonic and geotechnics of traveston crossing dam foundation. Proceed NZSOLDANCOLD Queenstown NZ 33(1):1–9

Berhane G (2010) Geological, geophysical and engineering geological investigation of leaky Micro-dam in the Northern Ethiopia. Agricult Eng Int CIGRE J 1346:XII

Rogers GD (2012) Foundation investigation at Hickory Creek Dam. Canton, Georgia

Kulkarni SK (1994) Engineering geological studies at Bhama irrigation project. Khed

Kulkarni SK et al (2007) Engineering geology in dam safety. In: ICOLD 75th annual meeting Saint Petersburg, Russo, pp 24–29

Gupte RB (1980) The Government of Maharashtra, P.W.D. handbook, chapter 6, part ii, engineering geology

IS 11315 (Part 12) (1992) Method for the quantitative description of discontinuities in the rock mass, Part 12: Drill core study

Deere DU, Deere DW (1988) The rock quality designation (RQD) index in practice. In: Kirkaldie L (ed) Rock classification system for engineering purposes, ASTM STP 984, Philadelphia, pp 91–101. https://doi.org/10.1520/STP48465S

Deere DU, Deere DW (1964) Technical description of rock cores for engineering purposes. Rock Mech Eng Geol 1(1):16–22

Camilo Quinones Rozo (2010) Lugeon test interpretation, revisited. In: Proceedings of the 30th annual USSD conference Sacramento—California, pp 405–414

IS 5529 (Part 2) (2006) Indian Standard in-situ permeability test- tests in bedrock

Goodman RE (2013) Some safety issues for dams on rock foundations. The Manuel Rocha Centennial lecture—at LNEC, Lisbon

Uromeihy A, Farrokhi R (2011) Evaluating groutability at the Kamal-Saleh Dam based on Lugeon tests. Bull Eng Geol Environ 64:382–387

IS 6066 (1994) Pressure grouting of rock foundations in. River valley projects recommendations (second revision)

Kamble RK (2011) Mapping seepage in the tailrace channel, Bhama Askhed dam: a case study. Bull Eng Geol Environ 70:643–649

Kulkarni SK (2006) Geological report-spillway and chute channel of Bhama Askhed

CWPRS, Pune (2008) Tracer studies for delineating path of seepage to the damaged portion of tail channel at Bhama Askhed irrigation project. Technical Report No. 4540 Applied Earth Sciences Division CWPRS, Pune, India

Kulkarni SK (2007) Engineering geological conditions leading to erosion of tail channels. Eval Exist Dams Act Plan Rest **I**:84–97

Aswar PB, Ullagaddi A (2022) 3-D Geomodelling for dam site characterization. In: Laishram B, Tawalare A (eds) Recent advancements in civil engineering. lecture notes in civil engineering, vol 172. Springer, Singapore. https://doi.org/10.1007/978-981-16-4396-5_71

Aswar DS (2018) Geomodelling for dam foundation stratum. http://hdl.handle.net/10603/265972

Seismic Soil-Structure Response of Plan Irregular Building on Piled Raft Foundation

C. L. Mahesh Kumar, T. Prathibha Reddy, and K. G. Shwetha

1 Introduction

An earthquake is caused by the shaking of the earth's surface, from which seismic waves are produced in the lithosphere as a result of the sudden release of energy from within the earth. Bedrock mass displacement is the cause of it. The ground will carry these seismic vibrations through it. The structure will be stimulated by the wave propagation in the soil, which will cause it to move in relation to the ground and produce diverse input motions. When the epicenter of high magnitude earthquake is located offshore, the seabed will get displaced to cause a tsunami. The earthquake will also trigger landslides and sometimes volcanic activity. Tectonic earthquakes will happen wherever adequate strain energy is stored to initiate rupture propagation along a fault plane. This energy will be released in a variety of ways, including seismic waves with radiated elastic strain, frictional heating of the fault surface, and rock-cracking earthquakes.

As per Indian Standard Code, plan irregularity is divided into two categories: (i) plan irregularity, i.e., torsional irregularities-entrant corners, diaphragm discontinuity, out-of-plane offset, and non-parallel system. (ii) Vertical irregularity, i.e., stiffness irregularity, mass irregularity, vertical geometric irregularity, in-plane discontinuity in vertical elements resisting lateral force, discontinuity in capacity. The current work investigates the impact of regular and irregular seismic soil response,

C. L. M. Kumar (✉) · K. G. Shwetha
Department of Civil Engineering, Nitte Meenakshi Institute of Technology, Yelahanka, Bangalore, Karnataka, India
e-mail: maheshkumar.cl@nmit.ac.in

K. G. Shwetha
e-mail: Shwetha.kg@nmit.ac.in

T. P. Reddy
Brindavan College of Engineering, Yelahanka, Bangalore 560063, India

K. K. Singh and C. S. Prasad Ojha (eds.), *Sustainable Management of Land, Water and Pollution of Built-up Area*, Society of Earth Scientists Series,
https://doi.org/10.1007/978-3-031-56176-4_5

and a comparison between regular and irregular types of structures has also been made. Soil structure interaction refers to the interaction between soil and structure for shell-like structures. In recent studies on review of piled-raft foundations for high-rise buildings, explored the application of piled-raft foundations, which combines elements of both pile foundations and raft foundations. With help of piled-raft analysis, studies showed that reduced settlements are seen. Consideration of SSI in loose soils is important for high rise buildings.

1.1 Foundation

The foundation is a structure which connects the superstructure to the ground and transfers the load from the structure to the ground. The foundation of a structure is always constructed below ground level to increase the lateral stability of structures.

The purpose of the foundation is:

- To increase the lateral stability of the structure as a whole.
- To prevent differential settlement.
- To secure a level and firm bed for building operations.
- To avoid lateral movement of supporting material.
- To distribute a structure load over a large bearing area to bring the intensity of loading within SBC.

The foundation should be designed:

- To transmit the load of the structure securely onto an adequate region of the soil so that the pressure induced in the soil are within limits.
- To guarantee no acceptability of structures i.e., the foundation region must be planned such that the central point of gravity (C.G.) of loads in the plan corresponds with the C.G. of the foundation region.

2 Methodology

In the present study, modelling and analysis of G + 15 storey regular and plan irregular (L and T) building has been done. The modeling and analysis of the building are carried out Using SAP2000 Software. As per I.S. 1893:2002, the earthquake loads are considered and seismic zone v is considered for the study. The type of analysis considered is the equivalent static method. A pile rafted foundation (Fig. 2) is considered for this analysis. Figure 2 shows un-deformed and deformed shape of rectangle building used in this study. Figures 3 and 4 show regular building with boundary condition with 3-layer soil and deformation of regular building with boundary condition with 3-layer soil, respectively.

The building models with soil-structure interaction (SSI) and without (SSI) are analyzed. The soil model was developed by using the finite element method. The soil

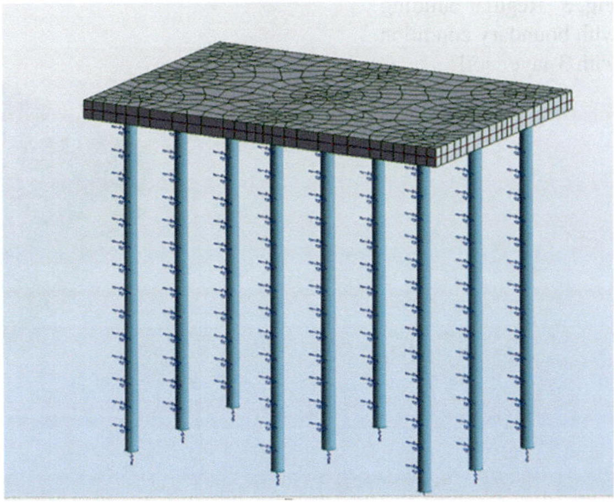

Fig. 1 Piled raft foundation (Anand N N et al. 2017)

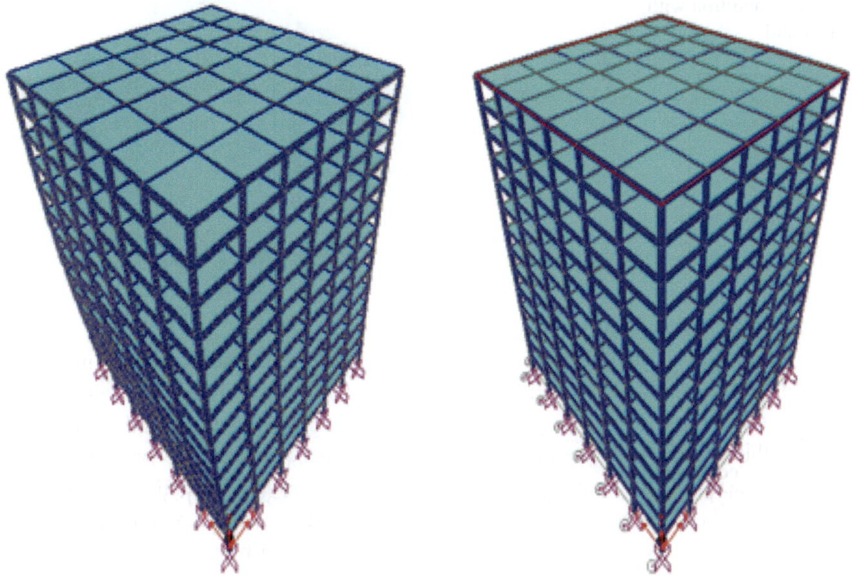

Fig. 2 Un-deformed and deformed shape of rectangle building

model is modeled by 3-layer, i.e., soft soil, stiff soil, soft rock, and is also modelled by 2-layer, i.e., soft soil, soft rock. The properties of different soils considered for analysis are presented in (Table 1).

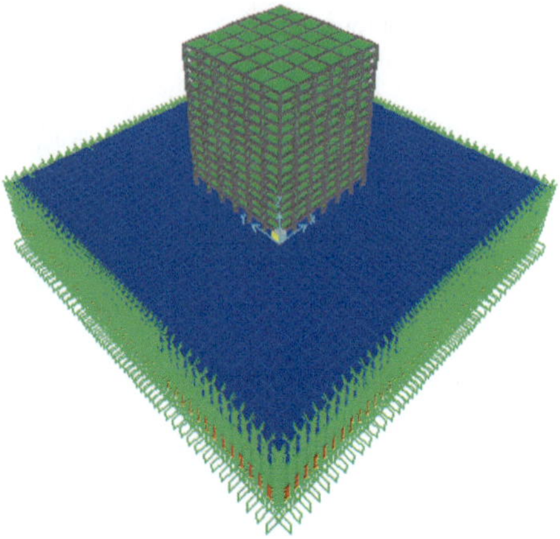

Fig. 3 Regular building with boundary condition with 3-layer soil

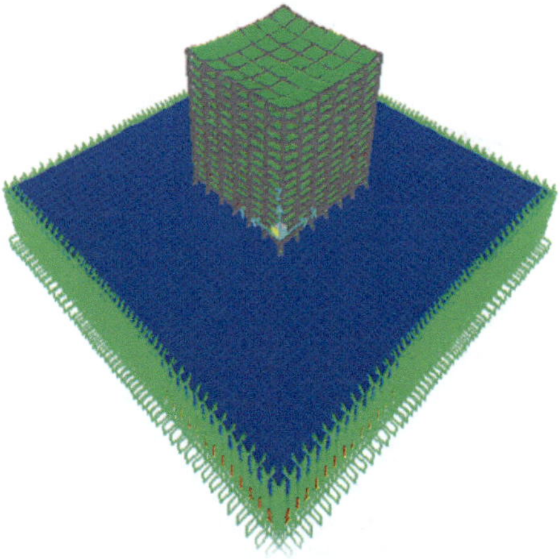

Fig. 4 Deformation of regular building with boundary condition with 3-layer soil

Table 1 Properties of different soils considered for study (B.S. 1377)

Type of soil	Properties of soil
Soft soil	Weight per unit volume = 15 kN/m^3 Modulus of elasticity (E) = 121,750 kN/m^2 Poisson's ratio = 0.3
Stiff soil	Weight per unit volume = 17 kN/m^3 Modulus of elasticity (E) = 3×10^5 kN/ m^2 Poisson's ratio = 0.3
Soft rock	Weight per unit volume = 22 kN/ m^3 Modulus of elasticity (E) = 1.5×10^6 kN/m^2 Poisson's ratio = 0.3

3 Results and Discussions

In this study, modelling and seismic analysis of bare frame regular and irregular buildings with different types of soil conditions, i.e., soft soil (Vs < 600 ft/s) and stiff soil (Vs = 600–1200 ft/s) and soft rock (Vs = 1200–2500 ft/s) for zone V was carried out by using SAP2000 software. Seismic parameters like lateral displacement, time period, frequency, and storey drift of the building are tabulated in Table 2. The results of the 3-layer soil model (soft, stiff, soft rock) and 2-layer soil model (soft soil, soft rock) and wind analysis was carried out for regular and irregular buildings for Zone V.

Figure 5 below shows the displacement variation for regular, L-shape, L-gap, and T-shape buildings without the effect of SSI. Table 3 shows displacement of buildings with SSI. For the T-shape, the displacement is 70.1% more compared to the L-gap model. When compared to L-gap displacement, the regular building displacement is 69.53% more. L shape displacement is 70.05% is more when compared to L-gap building.

The below Fig. 6 shows the variation of displacement for regular, L-shape, L-gap, and T-shape buildings with the effect of SSI. Table 4 shows displacement values for W_x. The displacement is 73.1% more for the L-shape than the L-gap model. When compared to regular building displacement, the T-shape displacement is 1.75% more.

Table 2 Displacement (mm) of all buildings without SSI

Number of storeys	Regular building	L-shape building	T-shape building	L-shape building with gap at the junction
15	302.6	307.9	309.3	92.2
14	298.4	301.8	303.2	90.2
13	292.1	293.8	295.3	87.7
12	283.4	283.7	285.2	84.5
11	272.3	271.2	272.6	80.6
10	258.6	256.4	257.8	76
9	242.5	239.2	240.5	70.8
8	224	219.7	221	64.9
7	203	198	199.1	58.4
6	179.6	174	175	51.2
5	153.8	147.9	148.7	43.4
4	125.6	119.7	120.3	35
3	95.2	89.6	90.1	26.1
2	63	58.2	58.5	16.9
1	30.5	27.2	27.4	7.9
0	0	0	0	0

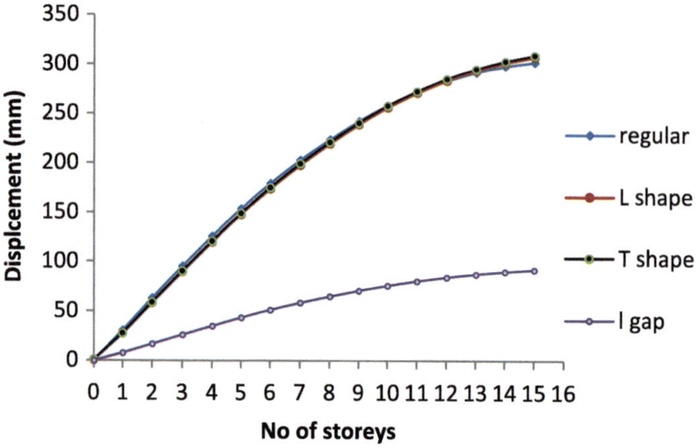

Fig. 5 Displacement versus No of storeys of building without SSI

Table 3 Displacement (mm) of buildings with SSI

Number of storeys	Regular building	L-shape building	T-shape building	L-shape building with gap at the junction
15	302.6	333.6	308	89.6
14	298.4	327.6	303.4	87.7
13	292.1	319.5	296.7	85.2
12	283.4	308.9	287.6	82.1
11	272.3	295.8	276.2	78.4
10	258.6	280.8	262.3	74.1
9	242.5	262.2	246	69.2
8	224	241.8	227.4	63.7
7	203	218.8	206.4	57.6
6	179.6	193.3	183	50.9
5	153.8	165.6	157.4	43.6
4	125.6	135.6	129.5	35.8
3	95.2	103.4	99.3	27.5
2	63	69.3	67.3	18.7
1	30.5	34.4	34	9.4
0	0	0	0	0

The least displacement is shown in the L-gap building when compared to the other buildings (Fig. 6).

Table 4 Displacement (mm) values for W_X-wind loading in X direction

Number of storeys	Regular building	L-shape building	T-shape building
15	22.7	57.7	60.4
14	22.4	56.8	59.4
13	22	55.4	58
12	21.4	53.6	56.1
11	20.5	51.3	53.7
10	19.5	48.5	50.8
9	18.3	45.3	47.4
8	16.9	41.5	43.5
7	15.2	37.5	38.2
6	13.4	32.8	34.4
5	11.5	27.8	29.2
4	9.3	22.5	23.6
3	7.1	16.8	17.1
2	4.7	11	11.5
1	2.3	5.2	5.4
0	0	0	0

Fig. 6 Displacement versus No of storeys of building with SSI

Wind Analysis

The below graph shows displacement versus the number of the storeys for wind analysis of wind speed 55 m/s. The T-shape building displacement is 64.41% more than a regular building and 4.4% more than an L-shape building. The least displacement is observed in a regular building (Fig. 7).

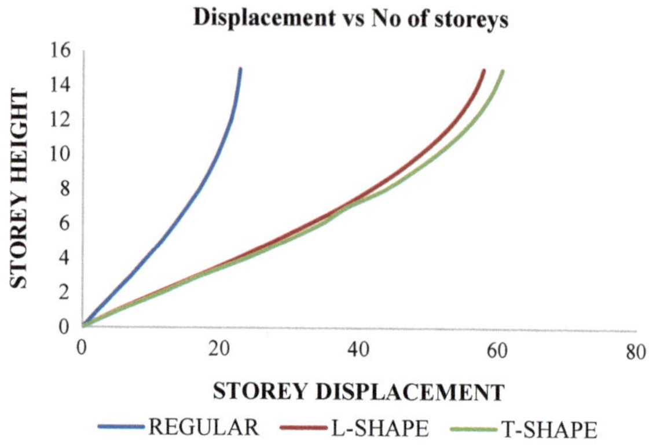

Fig. 7 Displacement versus storey height for different shapes

4 Conclusions

Case 1: Without Considering the Soil-Structure Interaction

1. **T section** shows more displacement when compared to all other sections.

Compared to L-shape building, T-shape building shows 1% more displacement while a regular building shows 9% more and L-gap building shows 70% more displacement.

Case 2: With Considering the Soil-Structure Interaction

2. **L section** shows more displacement compared to all other sections. i.e., compared to T-shape building, L-shape building shows 9% higher, and also shows 11% more than regular buildings and 69% more than L-gap building.

Case 3: Wind Analysis

i. T-section shows more displacement in the x-direction when compared to all other sections. i.e., 4.4% more displacement than the L-shape and 64.41% more than regular building
ii. L-section shows more displacement in the y-direction when compared to all other sections. i.e., 71.03% more displacement than T-shape building, which shows 62.82% more than a regular building.

Further Salient points are:

1. It is more important to consider soil-structure interaction for seismic evaluation of structure.
2. The displacement increases linearly in building models without considering the soil structure interaction.
3. Rectangular section (symmetrical structure) with soil-structure interaction shows 32% less displacement when compared to those without soil-structure interaction.
4. The 3-layered soil model shows less displacement when compared to 2-layered soil models by 9.1%.

5. T-section (axis-symmetrical) with soil-structure interaction shows 9% more displacement when compared to without soil-structure interaction. The displacement is 1% more in the 3-layered soil model when compared to the 2-layered soil model.
6. L-section (unsymmetrical) with soil-structure interaction shows 8.1% more displacement when compared without soil-structure interaction, and the 2-layered soil model shows 10% more displacement when compared to 3-layered soil model.

References

Anand N N, Payal P, Gunvant S (2017) A review on pile-raft foundation for high rise buildings. Int J Adv Eng Res Develop 4(11):322–326

B.S. 1377 (1990) Methods of test for soils for civil engineering purposes—Part 2: Classification tests and determination of geotechnical properties

Chaudhari RR, Kadam DK (2013) Effect of piled raft design on high-rise building considering soil structure interaction. Int J Sci Technol Res 2(6):72–79

Dabhi MG, Agrawal VV, Patel VB (2020) Soil structure interaction for basement system of multi storey building for different soil condition using static analysis in Etabs. SSRG Int J Civil Eng (SSRG-IJCE) 7(6)

Das B, Saha R, Haldar S (2016) Effect of in-situ variability of soil on seismic design of piled raft supported structure incorporating dynamic soil-structure-interaction. Soil Dyn Earthq Eng 84:251–268. https://doi.org/10.1016/j.soildyn.2016.02.015

Dutta SC, Das PK, Sengupta P (2017) Seismic behaviour of irregular structures. Struct Eng Int 27(4):526–545. https://doi.org/10.2749/222137917X14881938989765

Hokmabadi AS, Fatahi B (2016) Influence of foundation type on seismic performance of buildings considering soil–structure interaction. Int J Struct Stabil Dynam 16(08):1550043. https://doi.org/10.1142/S0219455415500431

Kumar MP, Naidu GT, Kumar TA (2015) Effect of soil-structure interaction on high rise R.C. regular frame structure with irregular bays subjected to seismic load. International J Res Eng Technol 4(10):1163–2319

Landi L, Conti F, Diotallevi PP (2015) Effectiveness of different distributions of viscous damping coefficients for the seismic retrofit of regular and irregular R.C. frames. Eng Struct 100:79–93. https://doi.org/10.1016/j.engstruct.2015.05.031

Lingaraju MKC, Girisha SK, Channabasappa SB, Karigowda M (2021) A study on dynamic behavior of natural draft cooling tower considering the effect of soil-structure interaction. Civ Environ Eng Rep 31(4), 17–32. https://doi.org/10.2478/ceer-2021-0047

Obulasetti A, Rao MVS, Kumar MKS (2016) Seismic analysis of R.C. framed office building considering soil structure interaction. Int J Technol Res Eng 4(4)

Patil S, Koraddi K (2020) Seismic response of shear wall buildings with rigid and flexible foundations. IUP J Struct Eng 13(1)

Reza Tabatabaiefar SH, Fatahia B, Samali B (2013) Lateral seismic response of building frames considering dynamic soil-structure interaction effects. Struct Eng Mech

Seismic Site Characterization and Ground Response Analysis for the Proposed Bridge Site at Prayagraj, India

Aakash Kumar and Saranika Das

1 Introduction

Every year, seismic hazards cause enormous damage all over the world. Seismic shaking at a particular location can vary in magnitude and duration depending on various factors like seismotectonic of the site, source parameter, path parameter, and other site-specific characteristics. Within a small area, the characteristics of seismic shaking may vary significantly due to the drastic change in soil conditions over short distances. For any construction project, in situ geotechnical and geophysical investigations are fundamental requirements to perform ground response analysis, seismic hazard studies, and liquefaction potential of any area. Due to the unavailability of geophysical data in the study area, an empirical correlation has been proposed using regression analysis (Anbazhagan et al. 2013; Chatterjee and Choudhury 2013; Hanumantharao and Ramana 2008; Imai and Yoshimura 1970). This proposed correlation is further used to obtain shear wave velocity (V_s) profile based on SPT-N number collected from boreholes (Kumar et al. 2018, 2022; Maheswari et al. 2010). These V_S values are used for seismic site characterization and ground response analysis (Naik and Choudhury 2014; Rao and Choudhury 2020; Sil and Haloi 2017). The long-span structure's behavior during an earthquake depends upon the earthquake motion, geotechnical characteristics of the underlying layer of soil, and path followed by seismic waves (Kumar and Debnath 2021; Sil and Haloi 2018). Various computer programs offered GRA like EERA, SHAKE, DEEPSOIL, and D-MOD2000. For

A. Kumar (✉)
National Institute of Technology, Meghalaya 793003, India
e-mail: p22ce002@nitm.ac.in; akash.mishra2312@gmail.com

S. Das
Indian Institute of Technology, Patna 801106, India
e-mail: saranika_1921ce13@iitp.ac.in

K. K. Singh and C. S. Prasad Ojha (eds.), *Sustainable Management of Land, Water and Pollution of Built-up Area*, Society of Earth Scientists Series,
https://doi.org/10.1007/978-3-031-56176-4_6

this study, 1D equivalent linear GRA has been performed using *DEEPSOIL v.7.0* software (NEHRP 2003).

The findings of this study will offer guidance for the earthquake-resistant design of structures, renovation of old structures, and seismic hazard analysis for Prayagraj, India.

2 Study Area Details

The proposed railway bridge is over the Yamuna river in the Prayagraj district of Uttar Pradesh state. This area is situated near the famous tourist place of Triveni Sangam. Prayagraj has zone factor of 0.16 (zone III), indicating a moderate seismic alert zone. The proposed bridge site with the positioning of boreholes is shown in Fig. 1.

Total of 22 borehole data has been collected at every pier location. In this project, drilling holes were carried out with Shell and Auger having 150 mm borehole diameter. Undisturbed samples have been collected at a suitable interval for the laboratory experiment. Subsoils are of good quality. Two layers of silty sand in the subsoil are present, a layer of loose silt and a layer of medium-dense to dense silt. After this, a very dense silty sand layer is found, continuing to the terminating depth of a few

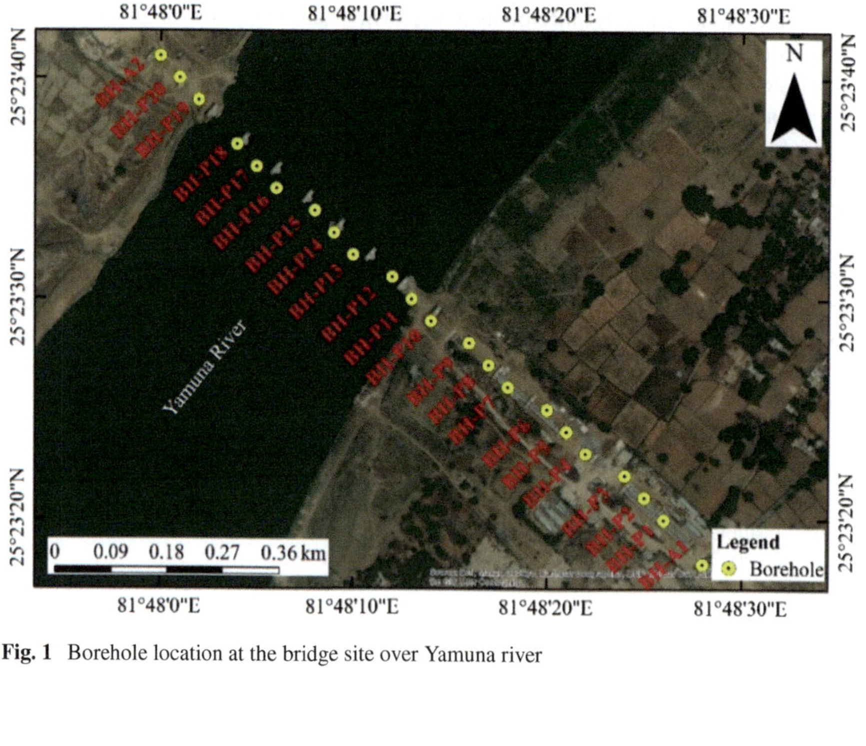

Fig. 1 Borehole location at the bridge site over Yamuna river

boreholes. With the study of boreholes, it has been observed that a hard layer of silty clay continued to the bottom.

3 Correlation Between V_S and SPT-N

The empirical correlation between SPT-N and V_S and has been developed for all types of soils with 438 SPT-N datasets obtained from 22 boreholes (Kumar et al. 2018, 2022). SPT-N used in this study has been taken from the borehole available for the bridge site. The proposed correlation has been developed by obtaining V_S from available correlations using weighted averaging techniques and non-linear regression equation (Chatterjee and Choudhury 2013; Hanumantharao and Ramana 2008). The proposed correlation has been given in Eq. 1. R^2 value for this correlation has been found 0.999, which shows that this correlation has a good prediction. In Fig. 2, the proposed correlation between SPT-N and V_S has been presented with $\pm$ 10% error bars.

$$V_S = 69.525N^{0.419} \tag{1}$$

Also, for validation purposes, present correlation has been compared with the correlations available for the Indian subcontinent and shown in Fig. 3.

In comparison with the existing correlations developed for the Indian subcontinent, the proposed correlation follows similar patterns and compares closely with the correlations available for Kolkata and Haryana regions (Chatterjee and Choudhury 2013; Rao and Choudhury 2020). Further, V_S profile obtained from the proposed correlation has been used for seismic site characterization and ground response analysis.

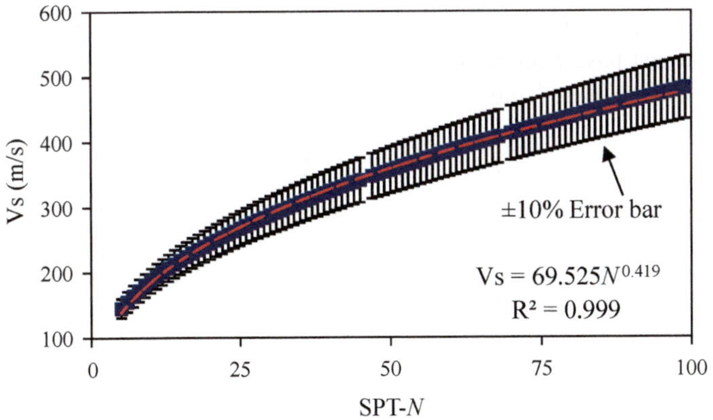

Fig. 2 Correlation between V_S and SPT-N

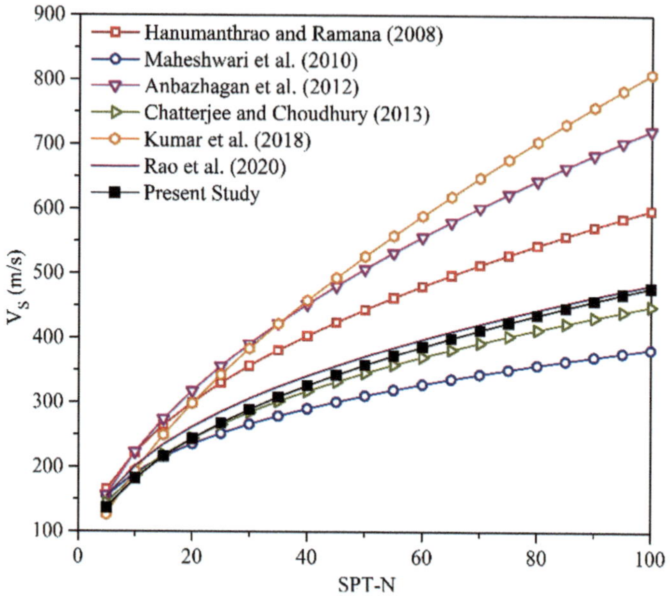

Fig. 3 Comparison of proposed correlation with correlation available for the Indian subcontinent

4 Seismic Site Characterization

Seismic site characterization serves more than to identify the proposed site conditions. It produces detailed information about the condition of a proposed bridge site so that valuable decisions can be made during construction phases and to keep this bridge safe during seismic hazards. It is essential to mitigate and identify the uncertainties of seismic hazards that may jeopardize the life of humans and the loss of property. SPT-*N* data from bore logs collected in the bridge site are analyzed in detail. A 30 m average SPT blow count (N^{30}) and average shear wave velocity ($V_S{}^{30}$) approach have been used in this study as per NEHRP guidelines (NEHRP 2003). Estimation of N^{30} and $V_S{}^{30}$ has been estimated by Eq. 2.

$$N^{30} = V_S^{30} = \frac{\sum_{i=1}^{n} d_i}{\sum_{i=1}^{n} \frac{d_i}{N_i or V_{Si}}} \tag{2}$$

The obtained $V_S{}^{30}$ varies from 250.5 to 413.04 m/s for the study site. Based on obtained $V_S{}^{30}$, Prayagraj region comes under site class *C* as per NEHRP guidelines (NEHRP 2003).

The estimated value of N^{30} and $V_S{}^{30}$ for each borehole location with site class has been displayed in Table 1.

Table 1 Seismic site classification based on NEHRP provisions

Borehole	Latitude	Longitude	N^{30}	V_S^{30} (m/s)	Site class
BH-A1	25° 23′ 18″	81° 48′ 28″	20.43	258.44	D
BH-P1	25° 23′ 20″	81° 48′ 26″	23.96	277.07	D
BH-P2	25° 23′ 21″	81° 48′ 25″	27.09	286.94	D
BH-P3	25° 23′ 22″	81° 48′ 24″	22.90	290.99	D
BH-P4	25° 23′ 23″	81° 48′ 22″	19.79	250.61	D
BH-P5	25° 23′ 24″	81° 48′ 21″	19.75	250.50	D
BH-P6	25° 23′ 25″	81° 48′ 20″	23.64	268.78	D
BH-P7	25° 23′ 26″	81° 48′ 18″	28.56	298.89	D
BH-P8	25° 23′ 27″	81° 48′ 17″	22.35	284.70	D
BH-P9	25° 23′ 28″	81° 48′ 16″	30.02	303.53	D
BH-P10	25° 23′ 29″	81° 48′ 14″	17.28	258.08	D
BH-P11	25° 23′ 30″	81° 48′ 13″	26.35	327.65	D
BH-P12	25° 23′ 31″	81° 48′ 12″	51.64	374.61	C
BH-P13	25° 23′ 32″	81° 48′ 10″	38.46	345.77	D
BH-P14	25° 23′ 33″	81° 48′ 9″	45.53	355.75	D
BH-P15	25° 23′ 34″	81° 48′ 8″	66.45	410.59	C
BH-P16	25° 23′ 35″	81° 48′ 6″	67.23	413.04	C
BH-P17	25° 23′ 36″	81° 48′ 5″	47.64	358.40	D
BH-P18	25° 23′ 37″	81° 48′ 4″	58.66	383.21	C
BH-P19	25° 23′ 39″	81° 48′ 2″	58.76	393.36	C
BH-P20	25° 23′ 40″	81° 48′ 1″	55.41	382.31	C
BH-A2	25° 23′ 41″	81° 48′ 0″	48.87	366.20	C

5 Ground Response Analysis

The main objective of GRA is to determine how local soils affect seismic waves by amplification or de-amplification and to estimate surface motion and seismic response spectra so that they can be utilized in the further design of the structure (Naik and Choudhury 2014; Sil and Haloi 2018). 1D equivalent linear GRA has been performed for five borehole locations (BH-A2, BH-P12, BH-P14, BH-P17, BH-P20) obtained from the proposed bridge site. For this analysis, (a) 2001-Bhuj earthquake, (b) 1991-Uttarkashi earthquake, and (c) 1999-Chi-Chi earthquake time history have been considered, as shown in Fig. 4.

In this analysis, the iterative approach has been used to estimate the linear damping ratio (ξ) and equivalent linear shear secant modulus (G) to approximate the non-linear hysteretic stress–strain property of the critically loaded soils. The inbuilt model has been used for the variation of G and ξ with shear strain (%) (Seed and Idriss 1970).

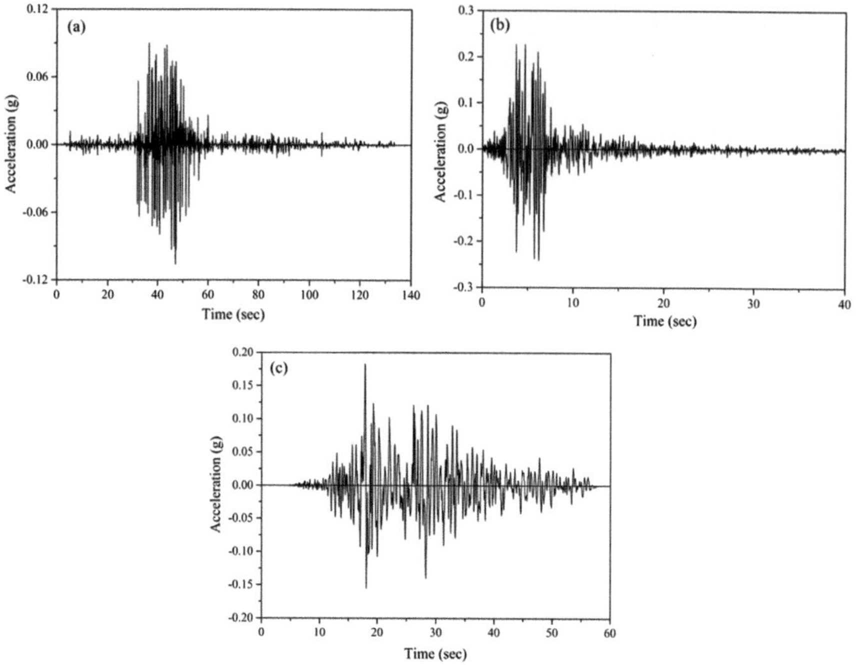

Fig. 4 A plot of the **a** 2001-Bhuj earthquake, **b** 1991-Uttarkashi earthquake, and **c** 1999-Chi-Chi earthquake time history considered for the present study

A detailed description of input parameters has been shown in Tables 2, 3, 4, 5 and 6 for five soil column models considered at selected borehole locations.

Soil level with SPT-*N* of more than 100 is considered a rigid bedrock level. This study estimates the bridge site's PGA and spectral acceleration (g) at the bedrock and surface level by considering local site effects. Surface acceleration time history

Table 2 Input parameters of BH-A2

Layer No.	Layer description	Thickness (m)	Unit weight (kN/m^3)	V_S (m/s)	Damping ratio
1	Clayey silt with sand mixture	6	18.93	285	0.5
2	Clayey silt with sand mixture	4.8	19.71	424	0.5
3	Clayey silt with sand mixture	9.7	19.71	417	0.5
4	Silty coarse-grained sand	9	17.66	447	0.5

Table 3 Input parameters of BH-P12

Layer No.	Layer description	Thickness (m)	Unit weight (kN/m^3)	V_S (m/s)	Damping ratio
1	Silty sand with mica	7.5	17.27	370	0.5
2	Silty coarse-grained sand	11.5	17.66	422	0.5
3	Silty sand with mica	3	17.27	355	0.5
4	Silty coarse-grained sand	6	17.66	450	0.5

Table 4 Input parameters of BH-P14

Layer No.	Layer description	Thickness (m)	Unit weight (kN/m^3)	V_S (m/s)	Damping ratio
1	Silty sand with mica	5	17.27	312	0.5
2	Silty sand with mica	8	17.27	285	0.5
3	Silty coarse-grained sand	6	17.66	458	0.5
4	Silty coarse-grained sand	12	17.66	434	0.5

Table 5 Input parameters of BH-P17

Layer No.	Layer description	Thickness (m)	Unit weight (kN/m^3)	V_S (m/s)	Damping ratio
1	Silty coarse-grained sand	6	17.66	340	0.5
2	Silty coarse-grained sand	10	17.66	432	0.5
3	Silty coarse-grained sand	6	17.66	379	0.5
4	Silty coarse-grained sand	10	17.66	392	0.5

Table 6 Input parameters of BH-P20

Layer No.	Layer description	Thickness (m)	Unit weight (kN/m^3)	V_S (m/s)	Damping ratio
1	Clayey silt with sand mixture	12.7	18.93	373	0.5
2	Clayey silt with sand mixture	8.75	18.93	392	0.5
3	Silty coarse-grained sand	7	17.66	463	0.5
4	Silty coarse-grained sand	7	17.66	463	0.5

with 2001-Bhuj earthquake time history has been depicted in Fig. 5. Due to internal dynamic properties of soil layers, input ground motion properties alter differently.

Variation of spectral acceleration (g) with period has been illustrated in Fig. 6 for selected borehole locations.

Fourier amplitude spectra for each location with their average have been illustrated in Fig. 7.

Out of the 22 boreholes at the bridge site, BH-P12, BH-P15, BH-P16, BH-P18, BH-P19, BH-P20, BH-A2 have been classified as class *C* and rest of boreholes as class *D*. Further, five boreholes have been selected for the 1D equivalent linear GRA. For each location, maximum displacement, effective stress, maximum PGA, average amplification factor, and maximum stress ratio are listed in Table 7.

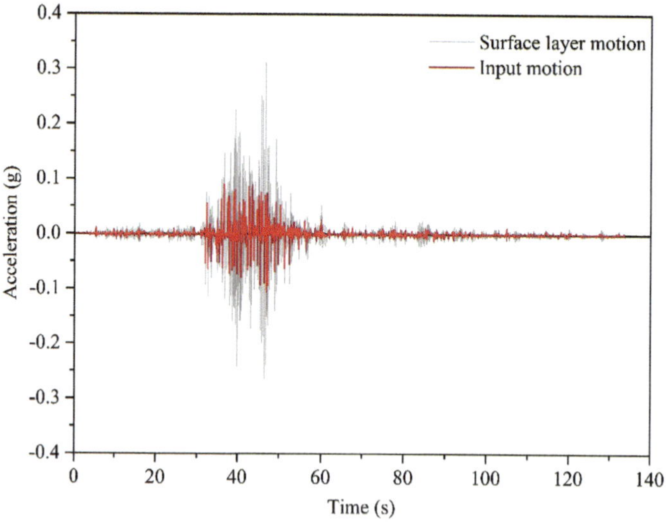

Fig. 5 Comparison of surface layer motion with bedrock motion for BH-A2

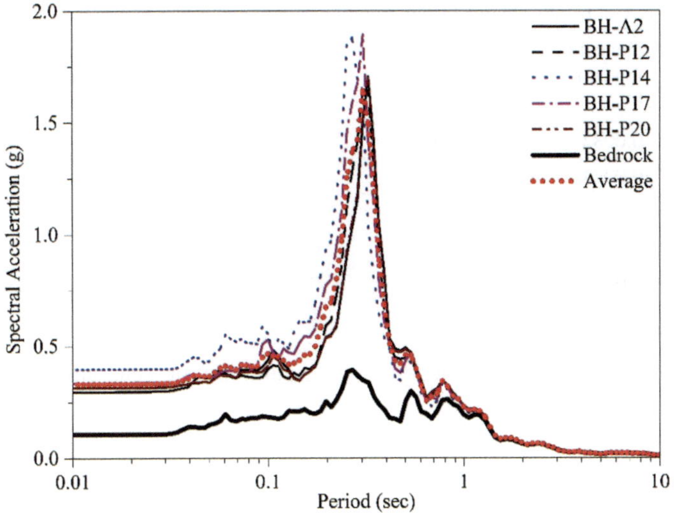

Fig. 6 Response spectra at surface and bedrock level with average response spectrum

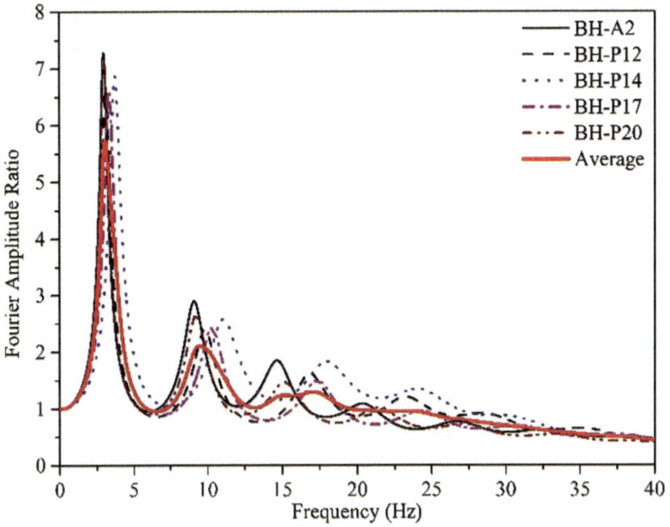

Fig. 7 Fourier amplitude spectra for each location with their average function

The maximum amplification factor has been found at BH-P14 as 3.92 and the minimum at BH-P12 as 3.30. Surface motion time history, response spectra, and Fourier amplification function have been generated for the proposed bridge site. These outcomes will benefit the seismic analysis of structures, mitigation of

Table 7 Important results obtained from ground response analysis

Borehole location	Deflection at top (m)	Effective stress (kPa)	Max. PGA (g)	Avg. amplification factor	Max. stress ratio
BH-A2	0.0082	328.648	0.31	3.42	0.31
BH-P12	0.0074	180.145	0.29	3.30	0.68
BH-P14	0.0068	112.51	0.39	3.92	0.83
BH-P17	0.0068	160.72	0.33	3.57	0.73
BH-P20	0.0081	355.097	0.31	3.38	0.30

earthquake induced hazards, and synthetic ground motion generation for the study area.

6 Conclusions

This study's main aim is to perform seismic site characterization and GRA for the proposed bridge site in Prayagraj, India. An empirical correlation between SPT-*N* and V_S has been established without geophysical data to estimate V_S profile using SPT-*N* obtained from 22 boreholes located at the proposed bridge site. Further, seismic site characterization of all sites has been done based on N^{30} and ${V_S}^{30}$ by following NEHRP guidelines.

References

Anbazhagan P, Kumar A, Sitharam TG (2013) Seismic site classification and correlation between standard penetration test N value and shear wave velocity for Lucknow City in Indo-Gangetic Basin. Pure Appl Geophys 170(3):299–318

Chatterjee K, Choudhury D (2013) Variations in shear wave velocity and soil site class in Kolkata city using regression and sensitivity analysis. Nat Hazards 69(3)

Hanumantharao C, Ramana GV (2008) Dynamic soil properties for microzonation of Delhi India. J Earth Syst Sci 117(2):719–730

Imai T, Yoshimura Y (1970) Elastic wave velocity and soil properties in soft soil. Tsuchito-Kiso 18(1):17–22

Kumar A, Debnath N (2021) Seismic behaviour of a typical rail bridge using north-east india specific synthetic ground motions under multi-support excitation. In: Das B, Barbhuiya S, Gupta R, Saha P (eds) Recent developments in sustainable infrastructure. lecture notes in civil engineering, vol 75. Springer, Singapore, pp 291–300

Kumar A, Harinarayan NH, Verma V, Anand S, Borah U, Bania M (2018) Seismic site classification and empirical correlation between standard penetration test N value and shear wave velocity for guwahati based on thorough subsoil investigation data. Pure Appl Geophys 175(8):2721–2738

Kumar A, Satyannarayana R, Rajesh BG (2022) Correlation between SPT-N and shear wave velocity (VS) and seismic site classification for Amaravati city India. J Appl Geophys 205:104757. https://doi.org/10.1016/j.jappgeo.2022.104757

Maheswari RU, Boominathan A, Dodagoudar GR (2010) Use of surface waves in statistical correlations of shear wave velocity and penetration resistance of Chennai soils. Geotech Geol Eng 28(2):119–137

Naik NP, Choudhury D (2014) Comparative study of seismic ground responses using DEEPSOIL, SHAKE, and D-MOD for soils of Goa, India. In: Geo-congress 2014: Geo-characterization and modeling for sustainability, pp 1101–1110

NEHRP (2003) NEHRP recommended provisions for seismic regulations for new buildings and other structures. FEMA 450

Rao VD, Choudhury D (2020) Estimation of shear wave velocity and seismic site characterization for new nuclear power plant region India. Nat Hazards Rev 21(4):06020004. https://doi.org/10.1061/(ASCE)NH.1527-6996.0000391

Seed HB, Idriss IM (1970) Soil moduli and damping factors for dynamic response analyses. EERC Rep (EERC 70–10) 1–15

Sil A, Haloi J (2017) Empirical correlations with standard penetration test (SPT)-N for estimating shear wave velocity applicable to any region. Int J Geosynth Ground Eng 3(3):1–13. https://doi.org/10.1007/s40891-017-0099-1

Sil A, Haloi J (2018) Site-specific ground response analysis of a proposed bridge site over Barak River along Silchar Bypass Road India. Innov Infrastruct Sol 3(1):1–19. https://doi.org/10.1007/s41062-018-0167-y

Sustainable Water Management

Flood Inundation Modeling: A Brief Review

Komal Vashist and K. K. Singh

1 Introduction

Information on the devastating events has a vital role in enhancing our design, planning, and construction processes in the food-risk region. These conclusions can be drawn from mathematical simulations of the spatial and temporal aspects of flood catastrophes. Mathematical models solve a set of governing equations for flood simulations and furnish precise details about the features of floods. The equations governing the flow of water are conservation of mass, momentum and energy and have form of nonlinear, partial differential equation. The solution of these equation requires to replace them with a set of statistical equation, which in turn need be solved with the aid of computers. A major advantage of flood simulation models is that alternative schemes for development or flood control can be quickly tested and compared.

Models were developed for flood hazard mapping, flood inundation modeling and flood risk assessment to compute the risk associated with the devastating events of various return periods (Farooq et al. 2019; Lindner and Miller 2012; Merwade 2017; Zhang et al. 2020). Some decades ago, the famous statistician George Box pointed out, 'All models are wrong, but some are useful'. In fact, although a model can never be perfect, models can be ranked, depending on the specific application, as very useful, somewhat useful or essentially useless (Burnham and Anderson, 2002). This ranking exercise is the essence of model selection (Laio et al. 2009) Generally, the bias tends to reduce as a model's dimension (or complexity) improves, however, the uncertainty tends to rise. There are models having less data requirement and are computationally cost-effective but have flaw of dropping out important features during simulation but on the other hand there are complex models that have

K. Vashist (✉) · K. K. Singh
Department of Civil Engineering, NIT Kurukshetra, Kurukshetra, India
e-mail: komal_61900039@nitkkr.ac.in

K. K. Singh and C. S. Prasad Ojha (eds.), *Sustainable Management of Land, Water and Pollution of Built-up Area*, Society of Earth Scientists Series,
https://doi.org/10.1007/978-3-031-56176-4_7

a high quality data requirements and high computational cost and difficulties in parameterization (Bates and Roo 2000).

It is impossible to cover all the research articles in one and summarize them but for this study a comprehensive review of various mathematical models available in literature has been done to evaluate further scope of action to improve modeling of flood affected areas. For this study a total of 42 research articles are reviewed from year 1995 to 2020. After the contribution of various researchers (Bates and Roo 2000; Horritt and Bates 2001; Horritt and Bates 2001) in this field, this area is getting importance day by day. Number of research articles published in various years in flood inundation area is shown in Fig. 1. In recent years with the advancement of technology numerous 1-D, 2-D and coupled1-D/2-D models were developed for riverine and urban flood modeling (Fan et al. 2017; Fewtrell et al. 2008; Yu and Lane 2006; Zainalfikry et al. 2020; Carrivick 2006; Seyoum et al. 2012; Timbadiya et al. 2015; Jacob et al. 2020), tabulated in Table 1 in the next section.

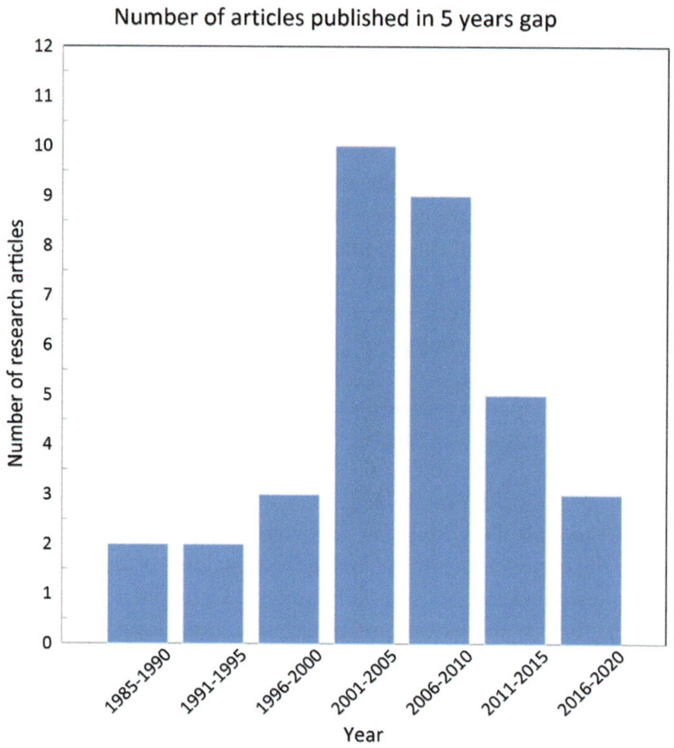

Fig. 1 Number of research articles published on flood at 5-year interval

Table 1 Models available in literature

Model name	Dimensionality	Numerical solution	Implicit/explicit
DIVAST	2D	Finite difference	Implicit
DELFT-3D	3D	Finite volume	Implicit/explicit
FESWMS	2-D	Finite element	–
HEC-RAS	1-D	Finite difference	Implicit
HYDRO-2E	2-D	Finite volume	Explicit
LISFLOOD-FP	coupled 1-D/2-D	Storage cell concept	
MIKE HYDRO (new version of MIKE 11)	1-D	Finite difference	Implicit
MIKE 21	2-D	Finite difference	Implicit
MIKE-FLOOD	Coupled 1-D/2D	Coupled solution of 1D/2D equations	Implicit
MIKE URBAN	Coupled 1-D/2D	Finite difference	–
RMA-2	2-D	Finite element	–
RUBER 20 Code	2-D	Finite volume	Explicit
SOBEK 1-D / 2-D	1-D/2-D	Finite difference	Implicit
TELEMAC 2-D	2-D	Finite element	Implicit/explicit
TELEMAC 3-D	3-D	Finite element or Finite volume	Implicit/explicit
TUFLOW	2-D	Finite difference	–

2 Governing Equation

De Saint Venant (1871) derived the equations for momentum and mass conservation that describe the unsteady open channel flow. The equations are

$$\frac{\partial Q}{\partial x} + \frac{\partial A}{\partial t} = q \tag{1}$$

$$\frac{\partial Q}{\partial t} + \frac{\partial \left[\Upsilon \frac{Q^2}{A} \right]}{\partial x} + gA\frac{\partial H_1}{\partial X} + \frac{gQ|Q|}{c^2 AR} = 0 \tag{2}$$

where: -

Q = Discharge (m^3/s)
A = Area (m^2)
q = Lateral inflow (m^2/s)
Υ = Momentum distribution coefficient
g = Gravity acceleration (m^2/s)
H_1 = Stage above the datum
c = Chezy's roughness coefficient_($m^{1/2}/s$)
R = Hydraulic radius (meter)

x = Distance in flow and
t = Elapsed time

In the current technical literature, there are three principal types of numerical solution methods: finite element method (FEM), method of characteristics, and finite difference method (FDM). The characteristics method was initially employed to solve the system of Saint–Venant equations but difficulties are brought on by the nonlinearity of the equations. This method was practically dismissed for non-equally spaced nodes, which are common in rivers. Finite element method determine the property of an element based on integral form of the governing equation whilst the FDM determines property of a single point/node and based on differential form of governing equation. Because computational fluid dynamics (CFD) problems typically demand discretization of the problem into a significant number of cells/grid points, the finite difference approach is used because the cost of solution prefers simpler, lower order estimates within each cell (Garcia and Kahawita 1986). The finite difference method now a day's stand over the other solving techniques in open channel modelling.

Numerical solution by method of finite difference can be explicit or implicit. Schemes which uses the nearby points data to calculate the flow variable at any location are explicit schemes. Whereas implicit schemes solve for all points concurrently at time t_{n+1}. Explicit numerical schemes are straightforward, but they necessitate tiny temporal increments. Implicit methods, which enable numerical solution over extensive time steps yet necessitate the solution of numerous continuous equations at each step due to their stability, as compared to other problems. The explicit approaches convert the Saint–Venant equations into a set of algebraic equations that are represented in space and time at each discretization point. One by one, these equations are solved. The implicit methods simultaneously solve the generated algebraic equations at all computing sites at a given time. One by one, these equations are solved. The implicit methods simultaneously solve the generated algebraic equations at all computing sites at a given time. Explicit schemes are helpful in conceptual explanation of numerical concepts and the best known explicit schemes are: the lax scheme, the Leap–Frog scheme. The Leap-Frog scheme was the earliest one ever used for numerical solution of one-dimensional wave equation. There is one explicit scheme which is widely used in flood routing i.e. Mac Cormack scheme, which is based on predictor and correction functions.

In the Preissmann scheme (Fig. 2) the value of an arbitrary function f_p (x, t) at point P can be estimated by

$$\frac{\partial f}{\partial t} = \left[\psi \frac{f_{j-1}^{n+1} - f_{j-1}^{n}}{\Delta t} + (1-\psi)\frac{f_{j}^{n+1} - f_{j}^{n}}{\Delta t}\right] \tag{3}$$

$$\frac{\partial f}{\partial x} = \left[(1-\theta)\frac{f_{j}^{n} - f_{j-1}^{n}}{\Delta x} + \theta \frac{f_{j}^{n+1} - f_{j-1}^{n+1}}{\Delta x}\right] \tag{4}$$

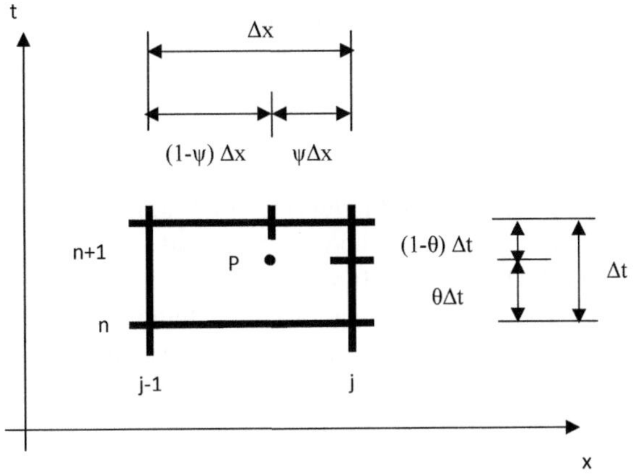

Fig. 2 Preissmann scheme

Where, n, n + 1, j-1, and j are the corresponding nodes for the point P, θ and ψ are the weighing coefficients.

When $\psi = 1/2$, the equation becomes Preissmann 4-point implicit scheme of finite difference or commonly known as box scheme. Some of the problems are solved using explicit scheme and some are solved by implicit schemes. Model with their description are tabulated in Table 1.

3 Literature Review

Model Review: Nowadays several numerical tools are available for modelling inundation process. These tools can be 1-D, 2-D, 3-D, and coupled 1D-2D models. In general, as a model's dimension (or complexity) rises, the bias tends to diminish, while the uncertainty tends to rise. As a result, a viable approach of 2D shallow water equations is the most physically realistic code that is used in practice. TELEMAC-2D is a 2-D model program that solves the 2-D SWE for a system of piecewise linear triangular finite elements in fractional steps (Galland et al. 1991), as is RUBER20, which solves 2D SWE using an explicit second-order finite volume (Haider et al. 2003). An example for model based on implicit finite difference scheme is MIKE 21 hydrodynamic model. There are variety of such models some based on FEM, some based on FDM which are available in literature (Nkwunonwo et al. 2020; Teng et al. 2017; Hunter et al. 2007). Models based on FEM, FDM, FVM are discussed below:

3.1 Finite Difference Based Methods

- *Implicit method*

Horritt and Bates 2002)—studied the performance of 1-D and 2-D models and found that there is difference in predictive performance of the models because each model give different response to chanage in friction parametres.

Yang and Rystedt 2002)—determine the model's stability issue, which is impacted by a variety of variables including the constraints, the bed resistance, the Courant number, and the flooding/drying conditions..

Haile and Tamiru Haile 2005)—said that obtaining a reliable simulation result from a 1-D model is a function of how accurately the model domain is represented. Study gives a good indication of magnitude of flood events and their features in the study region and implications of high resolutions DEM for flood modelling but the recent advances in high resolution topographic data acquisition encourages modelers to use and develop more complex models with the disadvantage of using excessive computational time and difficulty in defining the optimum parameter set.

Singh (2005)—the outcomes that were gathered for analysis were not as precise by the 1D model, but the model properly anticipated flooding along the whole course of the canal over time. 2D model is also tried but that shows run time error due to coarser resolution of DEM.

Werner et al. (2005)—the results show that if distributed observations of levels in the floodplain are used to constrain model uncertainties, in terms of anticipating distribution levels, the integrated 1D-2D strategy is somewhat better than the 1D approach, and the results show that flood inundation reliability can be greatly improved.

Lin et al. (2006)—discussed about the use of implicit 2D HD model to predict flood waves with abrupt surface gradients has been studied, and the results are comparable to those obtained using an explicit shock-capturing numerical model. Model estimates will not be significantly impacted by the advection parameters, but the flood wave speed will be significantly reduced if the advective and local acceleration terms are ignored.

Wright et al. (2008)—discerned that the 1-D models have a very short run time but are extremely susceptible to errors in the depiction of embankment heights and tend to underestimate the severity of inundation.

Cook and Merwade (2009)—asserted that because a 2D model's projections for various topography and geometric configurations vary less than a 1D model, the inundated extent it predicts is more accurate and real.

Patro et al. (2009)—found that calibration and validation findings of 1-D reveal that the model performs admirably in replicating river flow for the Mahanadi river basin's delta zone. The Manning's n values for the floodplain, rainfall at various sites., and elevations of the floodplain are included in the 2-D model. Although the pursuance of the coupled 1D-2D HD model in modelling flood inundation extent was determined to be good, differences in simulated water levels were noticed because structural features were not taken into account.

Kadam and Sen (2012)—concluded that the locations and patterns of flooding are found to be very similar, but the variation in flood inundation extent is due to coarser DEM resolution, lack of precise cross section placement, or embankment breaching by flood water. The most critical factor for a successful simulation of the connected model is a stable 1D and 2D model, and the computational time was reduced to keep the Courant number smaller than or equal to 1.

Villazón et al. (2013)—revealed that the overestimation of flooded areas and volumes using the 1-D model in combination with MIKE-GIS can be corrected by constraining the flooded surface area to given flood compartments.

Timbadiya et al. (2015)—revelaed from the sensitivity study that the floodplain had a lower response to Manning's roughness coefficient. The coefficient was found to work satisfactorily in a 1-D model created using surveyed cross sections. In some zones, the calibrated coupled 1D-2D model adequately estimated water levels, but the model's effectiveness was found to be unsatisfactory as a result of flaws in the data that were available for the investigations.

Balbhadra et al. (2018)—formed a rainfall-runoff model to assess the extent of a flooding provided a known amount of rainfall using HEC-HMS and HEC-RAS and found sensitive for curve number as compared to that of Manning's n and is not region sensitive.

Pramanik et al. (2010)—revealed from his study that in hydraulic models, in the absence of observed cross-section data, freely accessible SRTM_DEM-educed river cross-sections can be used to reproduce stage and discharge hydrographs with high precision.

- *Explicit methods*

Haider et al. (2003)—worked out that the ability of 2D algorithms to imitate flow around or through countless complications only described by their shape. It was demonstrated that utilising the code generates water levels that are near to measurements. One limitation is the cell size, which makes computing nearly impossible, and another is the interaction between close-by barriers, which causes waves and necessitates 3D developments that the 2D model is unable to accurately describe.

Mignot et al. (2006)—discovered that the event's numerical results have a large standard deviation, which represents the inadequacy of the topographical information provided, the ambiguity of the flood markers, and the regional impact of catastrophic events.

Lindner and Miller (2012)—offers a creative method to simulate the comprehensive set of flows when technological challenges make it difficult to monitor the flows in urban streams. While being bound by field-measured data, this alternate method produces valid rating curves using HD models. Additionally, this method has demonstrated the ability to replicate intricate stage-discharge correlations in small urban streams.

3.2 Finite Element Based Methods

Gee et al. (1990)—the results show that finite element techniques can accurately predict inundation in large-scale floodplain applications and are resistant to field variability in floodplain roughness assessment.

Horritt and Bates (2002)—find that the model cannot be used at its maximum level of resolution while sustaining numerical stability in most complicated topographic situations, and the model has affirmed the major findings in past studies, providing additional evidence of the suitability of this approach to large-scale floodplain modelling.

Bates et al. (1998), Bates et al. (1995)—observed that 2D FEM schemes have a lot of potential, especially in the realm of flood plain inundation extent, and investigations show that most aspects of widely accessible river flow data can be captured using a 2-D FEM with basic turbulence closure and a moderate density computational mesh.

Bates et al. (1997)—observed that the code incurs the lower computaional cost due to the use of a relatively long time step and raises the possibility that other aspects of the model parameterization other than boundary friction may also be important at particular spatial and temporal locations or for particular model predicted variables.

Horritt and Bates (2001), Horritt and Bates (2002)—the model produces the most complex calibration surface hence take time for calibration process and when calibrated against travel times and flooded areas, the results are comparable and independent of the calibration data.

Hunter et al. (2007)—the model is considered to yield optimum measures of fit that are roughly equivalent to level of skills, but that as flow grows, the model becomes less sensitive to Manning's n value and more sensitive to mesh resolution.

Cook and Merwade (2009)—employed a model developed by the Federal Highway Administration that used a finite element method to solve the depth-averaged average flow equation. The 2-D model's overall diversity in inundation extents for numerous topographic datasets and mesh resolution is lower than that of inundation extents predicted by the 1-D model.

3.3 Other Methods

- *Finite Volume:*

Robert (2001)—improved the accuracy of the digital terrain model would bring the largest improvement to the accuracy of the results, as per the two-dimensional modelling provided in the study. The study revealed that when the same common parameters were employed for the 1-D and 2-D models, the 1-D model's results were found to be less reliable than the 2-D model's.

Haider et al. (2003)—found that employing the code results in water levels that are comparable to measurements, proving the capacity of 2D codes to simulate flow through or over numerous hurdles that are merely represented by their geometry. One limitation is the cell size, which makes computing nearly impossible, and another is the interaction between nearby obstructions, which generates waves and necessitates 3D developments that cannot be directly epitomized by a 2D model.

Mignot et al. (2006)—find out that the numerical findings have a considerable standard deviation, which emphasizes the uncertainty in the flood markings, the inadequacy of geographic information furnished, and the local influence of disaster scenarios.

- *Storage cell concept:*

Bates and Roo (2000)—proved that the model's predictive capability for flood inundation scenarios is at its limit, and the model should be used within the uncertainty analysis framework to complement the level of accuracy.

Wright et al. (2008)—discovered that the differences between 1D and 2D models are of negligible relevance and that attention must be taken to incorporate multiple data sources for both model construction and validation when using remotely sensed data.

Horritt and Bates (2001), Horritt and Bates (2002)—concluded in their work that model is found simple to use and adequate for inundation prediction with resolution of model and water storage in low lying areas are found the important parameters for predicting the flood wave travel time but the model is significantly poorer when calibrated against flood wave travel time compared with when calibrated against inundated area.

4 Conclusion

In this study numerous mathematical models were studied for flood modeling. Although the 1-D models are computationally efficient but have the limitation of instability and fail to provide detailed information on flow. Studies have shown that 2D models can forecast factors like velocity, flood extent, and water level explicitly which are used to make flood risk management decisions. However, there are discrepancies in modeling findings, particularly for predictions based on 1D river to 2D floodplain linkages. From the review carried out on current flood modeling techniques, it can be perceived that coupled models are more preferable than other models and computationally efficient with less computational time and effort. Also for open channel modeling, models based on finite difference method are preferred over other numerical solution techniques.

References

Burnham KP (2002) Model selection and multimodel inference. Elsevier, New york

Balbhadra T, Ranjan P, Ajay K, Sajjad A, Ritu G (2018) Coupling HEC-RAS and HEC-HMS in precipitation runoff modelling and evaluating flood plain inundation map. 489

Bates PD, De Roo APJ (2000) A simple raster-based model for flood inundation simulation. J Hydrol 236(1–2):54–77. https://doi.org/10.1016/S0022-1694(00)00278-X

Bates PD, Anderson MG, Hervouet JM, Hawkes JC (1997) Investigating the behaviour of two-dimensional finite element models of compound channel flow. Earth Surf Process Landforms 22(1):3–17. https://doi.org/10.1002/(SICI)1096-9837(199701)22:1%3c3::AID-ESP667%3e3.0.CO;2-U

Bates PD, Horritt M, Hervouet JM (1998) Investigating two-dimensional, finite element predictions of floodplain inundation using fractal generated topography. Hydrol Process 12(8):1257–1277. https://doi.org/10.1002/(SICI)1099-1085(19980630)12:8%3c1257::AID-HYP672%3e3.0.CO;2-P

Bates PD, Anderson MG, Hervouet JM (1995) Initial comparison of two two-dimensional finite element codes for river flood simulation. Proc Inst Civ Eng Water Marit Energy 112(3):238–248. https://doi.org/10.1680/iwtme.1995.27886

By Robert JC, Cornel B, David JP (2001) Two-dimensional flood plain flow. ii: model validation. 6:406–415

Carrivick JL (2006) Application of 2D hydrodynamic modelling to high-magnitude outburst floods: an example from Kverkfjöll, Iceland. J Hydrol 321(1–4):187–199. https://doi.org/10.1016/j.jhydrol.2005.07.042

Cook A, Merwade V (2009) Effect of topographic data, geometric configuration and modeling approach on flood inundation mapping. J Hydrol 377(1–2):131–142. https://doi.org/10.1016/j.jhydrol.2009.08.015

Fan Y, Ao T, Yu H, Huang G, Li X (2017) A coupled 1D–2D hydrodynamic model for urban flood inundation. Adv Meteorol 2017:1–12. https://doi.org/10.1155/2017/2819308

Farooq M, Shafique M, Khattak MS (2019) Flood hazard assessment and mapping of River Swat using HEC-RAS 2D model and high-resolution 12-m TanDEM-X DEM (WorldDEM). Nat Hazards 97(2):477–492. https://doi.org/10.1007/s11069-019-03638-9

Fewtrell TJ, Bates PD, Horritt M, Hunter NM (2008) Evaluating the effect of scale in flood inundation modelling in urban environments. 5118:5107–5118. https://doi.org/10.1002/hyp

Galland J, Goutal N, Hervouet J (1991) TELEMAC : a new numerical model for solving shallow water equations. vol 14, pp 138–148

Garcia R, Kahawita RA (1986) Numerical solution of the St. Venant equations with the MacCormack finite-difference scheme. Int J Numer Methods Fluids 6(50):259–274.https://doi.org/10.1002/fld.1650060502

Gee DM, Anderson MG, Baird L (1990) Large-scale floodplain modelling. Earth Surf Process Landforms 15(6):513–523. https://doi.org/10.1002/esp.3290150604

Haider S, Paquier A, Morel R, Champagne JY (2003) Urban flood modelling using computational fluid dynamics. Proc Inst Civ Eng Water Marit Eng 156(2):129–135. https://doi.org/10.1680/wame.2003.156.2.129

Haile AT, Tamiru Haile A (2005) Integarting hydrodynamic models and high resolution DEM (LiDAR) for flood modelling. Itc 53(9):1689–1699. https://doi.org/10.1017/CBO9781107415324.004

Horritt S, Bates PD (2004) Evaluation of 1D and 2D numerical models for predicting river flood inundation. J Hydrol 268(1–4):87–99.https://doi.org/10.1016/S0022-1694(02)00121-X

Horritt MS, Bates PD (2001) Predicting floodplain inundation: raster-based modelling versus the finite-element approach. Hydrol Process 15(5):825–842. https://doi.org/10.1002/hyp.188

Horritt M, Bates P (2001) Effects of spatial resolution on a raster based model of flood flow. J Hydrol 253(1–4):239–249. https://doi.org/10.1016/S0022-1694(01)00490-5

Hunter NM, Bates PD, Horritt MS, Wilson MD (2007) Simple spatially-distributed models for predicting flood inundation: a review. Geomorphology 90(3–4):208–225. https://doi.org/10.1016/j.geomorph.2006.10.021

Jacob XK, Bisht DS, Chatterjee C, Raghuwanshi NS (2020) Hydrodynamic modeling for flood hazard assessment in a data scarce region: a case study of Bharathapuzha River Basin. Environ Model Assess 25(1):97–114. https://doi.org/10.1007/s10666-019-09664-y

Kadam P, Sen D (2012) Flood inundation simulation in Ajoy River using MIKE-FLOOD. ISH J Hydraul Eng 18(2):129–141. https://doi.org/10.1080/09715010.2012.695449

Laio F, Di Baldassarre G, Montanari A (2009) Model selection techniques for the frequency analysis of hydrological extremes. Water Resour Res 45(7):1–11. https://doi.org/10.1029/2007WR006666

Lin B, Wicks JM, Falconer RA, Adams K (2006) Integrating 1D and 2D hydrodynamic models for flood simulation. Proc Inst Civ Eng Water Manag 159(1):19–25. https://doi.org/10.1680/wama.2006.159.1.19

Lindner GA, Miller AJ (2012) Numerical modeling of stage-discharge relationships in urban streams. J Hydrol Eng 17(4):590–596. https://doi.org/10.1061/(ASCE)HE.1943-5584.0000459

Merwade ZLV (2019) Investigating the role of model structure and surface roughness in generating flood inundation extents using one- and two- dimensional hydraulic models. https://doi.org/10.1111/jfr3.12347

Mignot AP, Haider S (2006) Modeling floods in a dense urban area using 2D shallow water equations. J Hydrol 327(1–2):186–199.https://doi.org/10.1016/j.jhydrol.2005.11.026

Nkwunonwo UC, Whitworth M, Baily B (2020) A review of the current status of flood modelling for urban flood risk management in the developing countries. Sci African 7:e00269. https://doi.org/10.1016/j.sciaf.2020.e00269

Patro S, Chatterjee C, Mohanty S, Singh R, Raghuwanshi NS (2009) Flood inundation modeling using MIKE FLOOD and remote sensing data. J Indian Soc Rem Sens 37(1):107–118. https://doi.org/10.1007/s12524-009-0002-1

Pramanik N, Panda RK, Sen D (2010) One dimensional hydrodynamic modeling of river flow using DEM extracted river cross-sections. Water Resour Manag 24(5):835–852. https://doi.org/10.1007/s11269-009-9474-6

Seyoum D, Vojinovic Z, Price RK, Weesakul S (2012) Coupled 1D and noninertia 2D flood inundation model for simulation of urban flooding. J Hydraul Eng 138(1):23–34. https://doi.org/10.1061/(ASCE)HY.1943-7900.0000485

Singh S (2005) Analysis of uncertainties in digital elevation models in flood (Hydraulic) modelling. Geo-Information Sci 1–87

Teng J, Jakeman AJ, Vaze J, Croke BFW, Dutta D, Kim S (2017) Environmental Modelling and Software Flood inundation modelling: a review of methods, recent advances and uncertainty analysis. Environ Model Softw 90:201–216. https://doi.org/10.1016/j.envsoft.2017.01.006

Timbadiya PV, Patel PL, Porey PD (2015) A 1D–2D coupled hydrodynamic model for river flood prediction in a coastal urban floodplain. J Hydrol Eng 20(2):2267–2274. https://doi.org/10.1061/(ASCE)HE.1943-5584.0001029

Villazón L, Timbe T, Willems P (2013) "Comparative analysis of 1-D river flow models applied in a quasi 2-D approach for floodplain inundation prediction. MASKANA 4(1):107–126. https://doi.org/10.18537/mskn.04.01.08

Werner M, Blazkova S, Petr J (2005) Spatially distributed observations in constraining inundation modelling uncertainties. 3096:3081–3096. https://doi.org/10.1002/hyp.5833

Wright NG et al (2008) Case study of the use of remotely sensed data for modeling flood inundation on the river severn, UK. 134:533–540. https://doi.org/10.1061/(ASCE)0733-9429(2008)134

Yang X, Rystedt B (2002) Predicting flood inundation and risk using gis and hydrodynamic model : a case study at Eskilstuna, Sweden. Indian Cartogr 183–191

Yu D, Lane SN (2006) Urban fluvial flood modelling using a two-dimensional diffusion-wave treatment, part 1: mesh resolution effects. Hydrol Process 20(7):1541–1565. https://doi.org/10.1002/hyp.5935

Zainalfikry MK, Ab Ghani A, Zakaria NA, Chan NW (2020) HEC-RAS one-dimensional hydrodynamic modelling for recent major flood events in Pahang River. In: Lecture notes in civil engineering, vol 53, Springer, pp 1099–1115

Zhang Y, Zhou J, Lu C (2020) Integrated hydrologic and hydrodynamic models to improve flood simulation capability in the data-scarce three gorges reservoir region. Water 12(5):1462. https://doi.org/10.3390/w12051462

Prediction of Infiltration Rate of Rain Garden Using Conventional Models

Sandeep Kumar and K. K. Singh

1 Introduction

Urbanization is increasing day by day in India. This transition, which can bring India's urban population closer to 600 million by 2031, is not simply a shift of demographic change. Cities and towns are the center of the Report on Indian Urban Infrastructure and Services (Sivaramakrishnan 2015) depicting growth trajectory in India and this development spread is only possible by the propagation of concrete surfaces, roads, buildings roofs and parking lots over the watershed. Urban flooding is one of the most important issues for planners, policy makers and administrators. Human factors, hydrology and meteorology are responsible for urban flooding (Singh and Singh 2015). We can control human factors like improper maintenance of storm water drains, encroachment on water bodies, drains and inappropriate planning as shown in Fig. 1. Insufficient natural drainage and its blockage, extreme climate events and development in river flood plain are the major source of the urban flooding (Rafiq et al. 2016). All these surfaces cannot absorb, store or attenuate the rain water as normal forested lands do. To tackle the problems associated with storm water runoff, it is better to direct the runoff towards the ground water through infiltration.

The various techniques such as green-roof, bio retention, media filters, detention ponds, rain garden and infiltration surfaces, can increase the infiltration rate and are commonly called as best management practices (BMP) (Fletcher et al. 2015). Best Management Practice (BMP) is a natural infiltration concept that has been adopted by many industries and is attracting increasing interest (Davis 2008; Aaron et al. 2030). An approach to land management that addresses both water quality and urban hydrology by using natural elements like designed landscape features instead of conventional storm water management devices is growing in popularity

S. Kumar (✉) · K. K. Singh
Civil Engineering Department, NIT Kurukshetra, Kurukshetra, Haryana 136119, India
e-mail: sandeep_61900082@nitkkr.ac.in

K. K. Singh and C. S. Prasad Ojha (eds.), *Sustainable Management of Land, Water and Pollution of Built-up Area*, Society of Earth Scientists Series,
https://doi.org/10.1007/978-3-031-56176-4_8

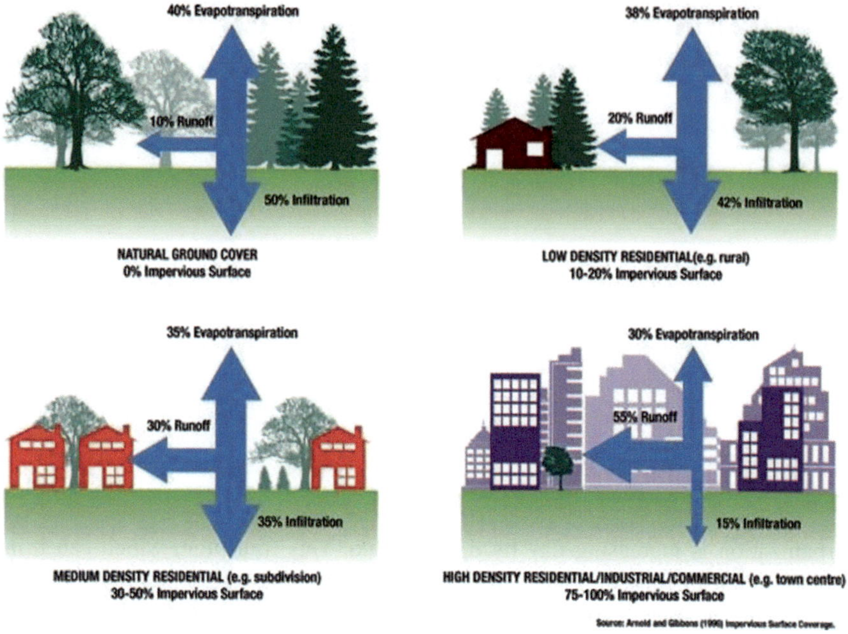

Fig. 1 Effects of imperviousness on runoff and infiltration

in American communities (Osheen and Singh 2020; Li and Davis 2008). Among all techniques, the rain garden technique is helpful in controlling the urban flooding and simultaneously helps in improving the water quality. As per hydrologic impact rain gardens were found more effective in both runoff volume and peak flow reductions as compare to other techniques (Yuan et al. 2017; Osheen and Singh 2019; Bhandari et al. 2018; Virahsawmy et al. 2014). Low impact development (LID) has been proposed as an alternative to conventional storm water management approaches. LID is an environmentally sensitive development method that focuses on preserving a hydrologically functional landscape after development (Penniman et al. 2013). The main advantage of LID is using natural processes to remove contaminants from water with out extra equipment, skilled labor or the energy as compared to other cleanup methods (Chang 2015). It is economical as compared to other techniques so it can be easily used by developing countries.

The hydrological performance of the rain garden relies on the type of soil (properties of soil such as permeability, filed capacity, organic matter content etc.) and vegetation type used in the rain garden. Retention and detention are two phrases used to describe storm water control. Evapotranspiration, plant interception, infiltration into the natural soil, and other retention activities are included (Yuan et al. 2017). Rain gardens are beneficial for lowering pollution, lowering peak runoff rates, and delaying runoff as a result (Davis et al. 2009; Dietz and Clausen 2005; Emerson and Traver 2008; Hunt and Lord 2004). Because of the high tolerability to the varying hydrological system of rain garden or bio-retention vegetation, they can be exploited in a broad diversification of surroundings (Weerasundara et al. 2016). It is generally known that the varied characteristics of rain gardens and their capacity to enhance quality by removing nutrients, oil and grease, heavy metals, and total suspended solids (TSS) from storm water flow (Davis et al. 2009; Dietz and Clausen 2005; Coffman et al. 1993). Rain garden is an attempt to "maximize all available physical, chemical, and biological pollutant removal processes found in the soil and plant region of a terrestrial forested community" (Davis et al. 2006). The results indicate that rain gardens are successful in reducing the impact of development on storm runoff (Li and Davis 2008). Time to the peak observed is also delayed by significantly with the help of rain gardens. In this study, the authors observed that the larger proportion of area for a rain garden gives promising results in controlling urban runoff (Li et al. 2016; Kumar 2019).

However, the efficiency of the rain garden depends widely on the available metrological and media records. Also the type of plant in the rain garden might affect its performance. In present work an attempt has been carried out to study the effect of plant on the efficiency of rain garden. Also the conventional techniques are employed to study the infiltration rate in rain garden as in certain conditions when it is not possible to carry out the experiment and at that time these conventional techniques can be used effectively.

2 Material and Methods

2.1 Experimental Setup

At the Hydraulics lab of the National Institute of Technology, Kurukshetra, India, four rain gardens were built. The two rain garden (RG) were of size 2 m × 1 m × 0.15 m (known as RG 1 and RG 2) and the other two are of size 1 m × 1 m × 0.05 m (called as RG 3 and RG 4) in RG 3 provision of overflow was made. Flatbed profiles might be seen on every rain garden. To keep the soil on the sides of the rain garden in place, the sides were encased with tiles or acrylic sheet. In Table 1 rain gardens size and plant used in the rain gardens is discussed.

To calculate the infiltration rate of RG1 and RG2, an equal amount of water (100 L) was applied in rain gardens through a cylindrical storage tank of capacity 170 L (through PVC pipes). Time to infiltration of a given volume of water is noted.

2.2 Materials

The properties of soil affect the hydraulic performance of a rain garden, that's why the testing of the soil is necessary. In order to get more information about the properties of the soil in situ, some additional tests such as the Atterberg limit, particle size

Table 1 Details of the size of rain garden and number of plants grown

Name (Size of RG)	Type and number of plants
RG 1(2 m × 1 m × 0.15 m)	Scutch grass = 300 Candytuft flower = 220 Daisy flower = 24
RG 1(2 m × 1 m × 0.15 m)	Scutch grass = 240 Candytuft flower = 110 Daisy flower = 12
RG 3 (1 m × 1 m × 0.05 m)	Scutch grass = 148 Marigold flower = 9 Daisy flower = 12
RG 4 (1 m × 1 m × 0.05 m)	Scutch grass = 140 Marigold flower = 9 Daisy flower = 12

distribution were performed. Sand (75–425 micro metre) and clay (2 micro metre), each making up 55.41% and 44.59% of the soil in the garden, are the predominant components. Results shows that the soil is brown in color and classified as a sandy loam. Liquid limit, plastic limit, and plasticity index were the Atterberg limits, and their respective values were 24.5, 16.00, and 8.50%.

Four types of vegetation were used for calculating infiltration rate (scutch grass, candytuft flower, marigold flower and daisy flower). Rate of infiltration was also checked on plant height (plant density) and type of plant used. Large amount of storm water when enters into the rain garden that case was consider as an overflow case. Overflow case is studied in RG3 for calculating infiltration rate, an outlet is provided for collecting outflow in a separate container. RG4 has no outlet. All the four vegetations (Scutch Grass, candytuft flowers, marigold flower and daisy flower) were used in rain garden during experiment with same number of plants. The first inflow rate was 7 L/min. after that the second inflow rate to the rain garden was 5.7 L/min. Table 2 gives range of various data collected from rain gardens.

Table 2 Range of information gathered from rain gardens

Variables	Temperature (degree Celsius)	Water content (%)	Inflow to RG (L/min)	Plant height (cm)	Number of plant	Water depths (cm)	Time elapsed in drop of water depth (seconds)	Infiltration rate (cm/h)
Min	11	4.36	5.4	0	0	0	366	0
Max	34	15.1	7	34	300	5	3752	12.46
Mean	20.41	10.27	6.33	16.81	75.38	1.59	1147.02	5.03
SD	7.29	3.40	0.79	13.42	92.80	1.37	693.08	2.65

Table 3 Value of a and b used for prediction the infiltration rate in Kostiakov model

Plant Name	a	b
Scutch Grass	5.269	– 0.207
Daisy Flower	5.013	0.084
Marigold Flower	4.266	0.642
Bare Soil	5.032	0.182

2.3 Data Set

The study area's geographic coordinates are 29.9490° N and 76.8173° E. For the purpose of predicting the infiltration rate (f(t)), the effects of temperature (t), plant height (H), initial moisture content (wc), plant count (N), water depth (d), input to the rain garden (R), and time (T) are investigated.

3 Modeling Techniques

3.1 Kostiakov Model (KM)

The Kostiakov model (Subramanya 2020) is described in detail as follows:

$$f(t) = at^b \tag{1}$$

where a and b are constants and t = time elapsed in the drop of water depth in hr. The values of a and b adopted during the present study are reported in Table 3.

3.2 Multi-linear Regression (MLR)

For establishing the relationship between dependent and independent variables, the MLR approach, which is based on linear regression, is utilized. The following is the formula for MLR that is frequently used: (Sihag et al. 2021)

$$f(t) = c_0 + c_1x_1 + c_2x_2 + c_3x_3 + \cdots + c_nx_n \tag{2}$$

In Eq. (2), f(t) stands for the dependent variable (output or infiltration rate), c_0, $c_1, c_2, c_3, \ldots, c_n$ stands for the regression coefficients, and $x_1, x_2, x_3, \ldots, x_n$ stands for the independent variables.

The variables f(t) with r, H, N, and T are used in the multi-linear regression analysis. The multi-linear equation for plants has four explanatory variables, one of which is bare soil. The following multi-linear models were created:

Equation (3) for grass plant of MLR

$$f(t) = 4.469 + 0.0443 * R + 9.394e{-}02 * H + 1.496e{-}02 * N - 2.713 * T \quad (3)$$

Equation (4) for candytuft flower plant of MLR

$$f(t) = -17.467 + 2.065 * R + 6.076e{-}03 * H + 2.210e{-}02 * N + 6.848 * T \quad (4)$$

Equation (5) for Daisy flower plant of MLR

$$f(t) = -0.994 - 4.112e{-}02 * R + 0.1762 * H + 0.1256 * N - 1.370 * T \quad (5)$$

Equation (6) for Marigold flower plant of MLR

$$f(t) = 12.779 - 0.536 * R - 4.732 * T \quad (6)$$

Equation (7) for Bare soil of MLR

$$f(t) = 4.761 + 0.388 * R - 2.204 * T \quad (7)$$

4 Model Performance

Performance of Kostiakov model and MLR model were evaluated by statistical parameters i.e. correlation coefficients (CC) and root mean square error (RMSE). The relationships used in study are given below in Eqs. (8) and (9)

Correlation Coefficient (CC): The value of CC is calculated as:

$$CC = \frac{m\sum_{i=1}^{m} o_i t_i - \left(\sum_{i=1}^{m} o_i\right)\left(\sum_{i=1}^{m} t_i\right)}{\sqrt{m\left(\sum_{i=1}^{m} o_i^2\right) - \left(\sum_{i=1}^{m} o_i\right)^2}\sqrt{m\left(\sum_{i=1}^{m} o_i^2\right) - \left(\sum_{i=1}^{m} t_i\right)^2}} \quad (8)$$

Root-mean-square error (RMSE): The value of RMSE is calculated as:

$$RMSE = \sqrt{\frac{1}{m}\left(\sum_{i=1}^{m}(o_i - t_i)\right)^2} \quad (9)$$

where, o_i = observed rate of infiltration, t_i = predicted rate of infiltration and m = number of observations.

5 Results and Discussion

As shown in Fig. 2, It has been noticed that the rain garden with higher number and height of plants in each type of vegetation are having the higher infiltration rates. And overall study has shown that infiltration rates varies as – grass plants > chandni flower plants > marigold flower plants > daisy flower plants.

So after overall analysis, we can say that rain gardens having the maximum grass proportion are the best low-impact development techniques and infiltrate the water most expeditiously.

5.1 Performance of Kostiakov Model (KM) and MLR Model

Infiltration rate is predicted using KM and MLR for a different type of plants in rain garden. The predicted infiltration rate vs the actual infiltration rate for different plant types are reported in Fig. 3 in form of agreement diagram. Figure 3 shows the agreement diagram between actual and predicted infiltration rate for KM and MLR. For Scutch Grass, Candytuft flower, Daisy Flower plant MLR predict better behavior as compared to KM as the corresponding values for MLR lies near the agreement line. Most of the values of MLR lies in the range of ± 25% error line. Most of the values are on perfect agreement line of MLR as compared to KM. From Table 4 it can be inferred that the choice of technique or model adopted widely depends on the plant type. In case of marigold, KM perform much better than MLR, whereas in case of Scutch Grass, daisy flower and Candytuft flower MLR perform better than KM.

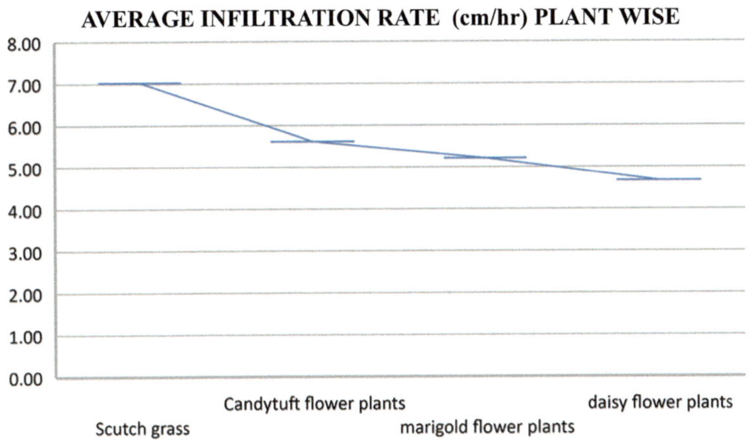

Fig. 2 Observed average infiltration rate plant wise

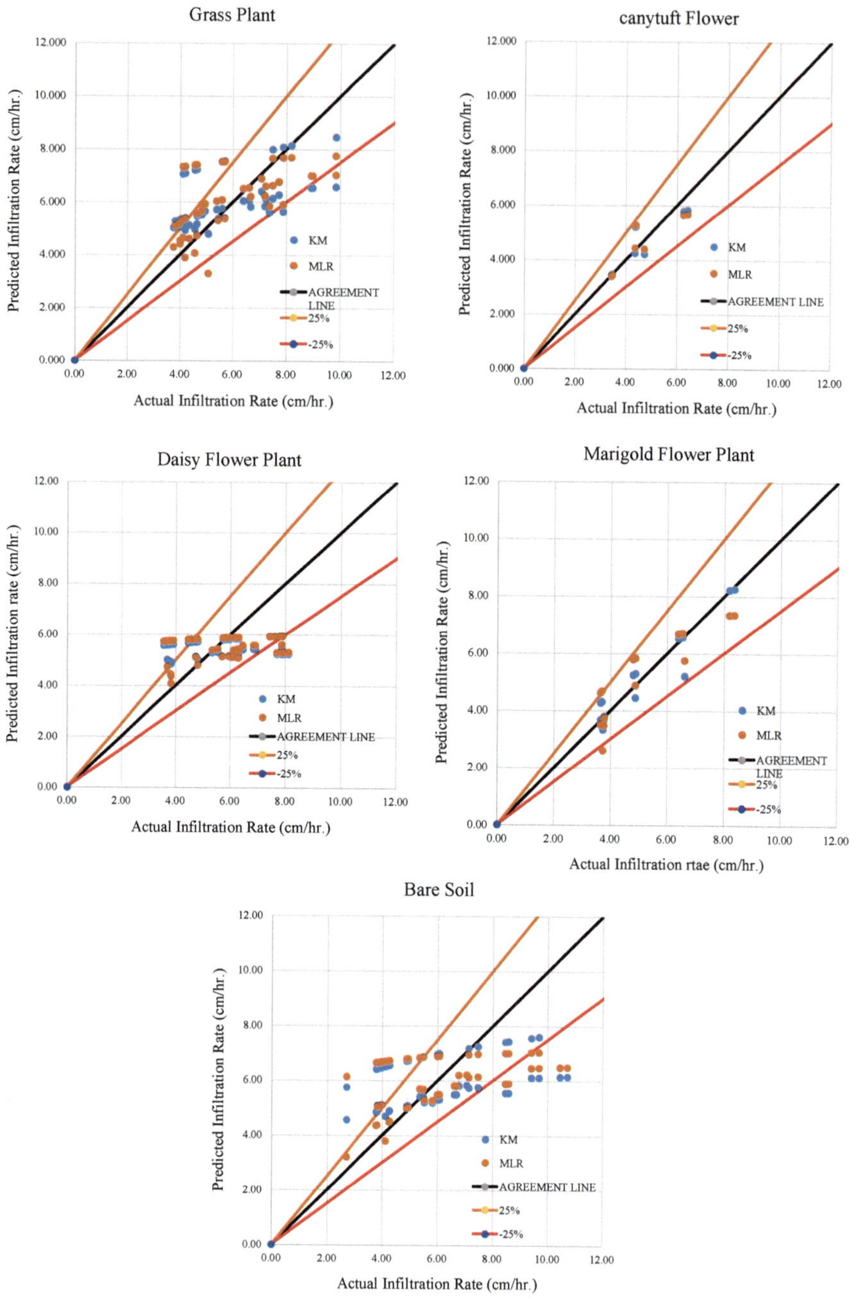

Fig. 3 Agreement diagram between Kostiakov model, MLR and observed infiltration rate for Scutch Grass, Candytuft flower, Daisy Flower, Marigold Flower Plant and Bare Soil for all the RG

Table 4 Comparison of different modelling techniques

Plant Name	Kostiakov model		MLR	
	CC	RMSE	CC	RMSE
Scutch Grass	0.473	1.974	0.663	1.741
Candytuft	0.856	0.587	0.946	0.413
Daisy	0.209	1.481	0.996	0.115
Marigold	0.952	0.538	0.918	0.735
Bare	0.407	1.958	0.471	1.912

6 Conclusion and Future Scope

In present work, infiltration capacity of the soil has been studied for different types of vegetation planted over it. Infiltration has been determined experimentally and conventional models (KM & MLR). Following are the points to be noted during the present study.

1. As shown in Fig. 2, Scutch grass shows maximum infiltration rate whereas, daisy flower plant shows the minimum infiltration rate. The order of decreasing infiltration rate is Scutch Grass plants > Candytuft flower plants > Marigold flower plants > Daisy flower plants.
2. The accuracy of conventional model depends on the type of vegetation. Marigold flower plant gives better prediction using KM whereas scutch grass, daisy and candytuft gives better results using MLR.
3. From the above discussion it can be concluded that the conventional model MLR is performing better than KM for predicting infiltration rate (Table 4).

Future Scope

This study can be further used on the:

1. Quantification of infiltration and evapotranspiration from a rain garden to properly assess its volume reduction potential.
2. Rain garden's vegetation and soil may act as filter media for removing pollutants from storm water. So, Pollutant removal efficiency of rain garden could be studied.
3. The permeability and water capacity of the rain garden will be dictated by the planting combination (i.e., the proportion of compost and sand), allowing the impact of planting mixes on rain gardens performance and design to be assessed.

Acknowledgements Sandeep Kumar, the first author, expresses gratitude to MHRD, GOI for funding the current work with a Ph.D. scholarship award (2K19/NITK/PHD/61900082).

References

Aaron B, Gauri K, Jessica P, Kyle T (2012) Redesigning the urban water cycle a vision for Redgedale mall 2030. (Online). Available: https://conservancy.umn.edu/handle/11299/185362

Bhandari S, Jobe A, Thakur B, Kalra A, Ahmad S (2018) Flood damage reduction in urban areas with use of low impact development designs. In: World environmental and water resources congress 2018, May 2018, vol 1, no i, pp 52–61. https://doi.org/10.1061/9780784481431.006

Chang CWGV (2015) Reimagining urban drainage in the world's biggest construction sites: three LID stories in eastern China. In: Proceedings of 2015 international low impact development conference 2015—LID it work in all climates soils, pp 384–393. https://doi.org/10.1061/978 0784479025.039.

Coffman L, Green R, Clar M, Bitter S (1993) Development of bioretention practices for stormwater management. Water Resour Plan Manag Urban Water Resour 6062:126–129. https://doi.org/10. 14796/jwmm.r176-02

Davis AP (2008) Field performance of bioretention: hydrology impacts. J Hydrol Eng 13(2):90–95. https://doi.org/10.1061/(asce)1084-0699

Davis AP, Shokouhian M, Sharma H, Minami C (2006) Water quality improvement through bioretention media: nitrogen and phosphorus removal. Water Environ Res 78(3):284–293. https:// doi.org/10.2175/106143005x94376

Davis AP, Hunt WF, Traver RG, Clar M (2009) Bioretention technology: overview of current practice and future needs. J Environ Eng 135(3):109–117. https://doi.org/10.1061/(asce)0733-9372

Dietz ME, Clausen JC (2005) A field evaluation of rain garden flow and pollutant treatment. Water Air Soil Pollut 167(1–4):123–138. https://doi.org/10.1007/s11270-005-8266-8

Emerson CH, Traver RG (2008) Multiyear and seasonal variation of infiltration from storm-water best management practices. J Irrig Drain Eng 134(5):598–605. https://doi.org/10.1061/(asc e)0733-9437

Fletcher TD et al (2015) SUDS, LID, BMPs, WSUD and more—the evolution and application of terminology surrounding urban drainage. Urban Water J 12(7):525–542. https://doi.org/10. 1080/1573062X.2014.916314

Hunt W, Lord W (2004) Bioretention performance, design , construction , and maintenance. North Carolina Coop Ext Serv

Kumar A (2019) Experimental study of a rain garden. National Institute of Technology, Kurukshetra

Li H, Davis AP (2008) Urban particle capture in bioretention media. II: theory and model development. J Environ Eng 134(6):419–432. https://doi.org/10.1061/(asce)0733-9372(2008)134: 6(419)

Li J, Li Y, Li Y (2016) SWMM-based evaluation of the effect of rain gardens on urbanized areas. Environ Earth Sci 75(1):1–14. https://doi.org/10.1007/s12665-015-4807-7

Osheen and K. K. Singh (2019) Rain garden—a solution to urban flooding: a review, vol 30. Springer, Singapore

Osheen and Singh KK (2020) The influence of slope profile on rain gardens' hydrological performance. In: World environmental and water resources congress 2020, May 2020, no Gallagher 2011, pp 145–153. https://doi.org/10.1061/9780784482988.015

Penniman D, Hostetler M, Borisova T, Acomb G (2013) Capital cost comparisons between low impact development (LID) and conventional storm water management systems in Florida. Suburb Sustain 1(2). https://doi.org/10.5038/2164-0866.1.2.1

Rafiq S, Salim R, Nielsen I (2016) Urbanization, openness, emissions, and energy intensity: a study of increasingly urbanized emerging economies. Energy Econ 56:20–28. https://doi.org/10.1016/ j.eneco.2016.02.007

Sihag P, Dursun OF, Sammen SS, Malik A, Chauhan A (2021) Prediction of aeration efficiency of Parshall and modified Venturi flumes: application of soft computing versus regression models. Water Supply 00:1–18. https://doi.org/10.2166/ws.2021.161

Singh RB, Singh S (2011) Rapid urbanization and induced flood risk in Noida, India, no July 2015. https://doi.org/10.1080/10225706.2011.629417

Sivaramakrishnan KC (2015) Infrastructure and services. Gov. Megacities 56–81. https://doi.org/10.1093/acprof:oso/9780199454136.003.0004

Subramanya K (2020) Engineering hydrology, 5th edn. Tata Mcgraw Hill Education Limited, New Delhi, India

Virahsawmy HK, Stewardson MJ, Vietz G, Fletcher TD (2014) Factors that affect the hydraulic performance of raingardens: Implications for design and maintenance. Water Sci Technol 69(5):982–988. https://doi.org/10.2166/wst.2013.809

Weerasundara L, Nupearachchi CN, Kumarathilaka P, Seshadri B, Bolan N, Vithanage M (2016) Bio-retention systems for storm water treatment and management in urban systems. https://doi.org/10.1007/978-3-319-41811-7

Yuan J, Dunnett N, Stovin V (2017) The influence of vegetation on rain garden hydrological performance. Urban Water J 14(10):1083–1089. https://doi.org/10.1080/1573062X.2017.1363251

Macropore Flow and Its Characteristics

Anoop Pandey and Richa Ojha

1 Introduction

A significant number of agricultural and environmental problems involve understanding of water and contaminant transport through the fractured vadose zone. Here, the vadose zone refers to the unsaturated zone spanning from the land surface to the first unconfined aquifer. A key component in these studies is the development of flow and transport models which approximately represent the natural systems. When addressing a physical problem, it is a common practice to begin by studying the problem for a simplified physical domain and under controlled boundary conditions. Water movement studies in the vadose zone have evolved in the same manner. Darcy (1856) studied the fluid flow in repacked sand columns and provided a simplistic approach to model the steady saturated flow through homogeneous and isotropic soil media. Some practical solutions of Darcy's law were provided by Schlichter (1899). To the end of the nineteenth century flow through porous media was constrained only to the saturated flow domain. Buckingham (1907) extended Darcy's law to flow in unsaturated media by including the capillary-induced pressure in the hydraulic potential term. In 1931, Richards derived the governing equation for unsteady flow through unsaturated soil media by combining the mass conservation of water with the Darcy-Buckingham equation.

Remarkably, although Richards laid the foundation for modelling flow through variably saturated porous media with the assumption of equilibrium flow of water, Haines (1930) discovered that unsaturated porous media can have different moisture content values at a particular capillary potential. Haines was the first who discovered

A. Pandey (✉) · R. Ojha
Indian Institute of Technology Kanpur, Kanpur 208016, India
e-mail: anoop20@iitk.ac.in

R. Ojha
e-mail: richao@iitk.ac.in

K. K. Singh and C. S. Prasad Ojha (eds.), *Sustainable Management of Land, Water and Pollution of Built-up Area*, Society of Earth Scientists Series,
https://doi.org/10.1007/978-3-031-56176-4_9

how a single pore (pore body with the neck at the top and bottom) affects the dynamics of water filling and draining. This effect is also known as the hysteresis behavior of soil water. The assumption that was carried from Darcy to Richards about the homogeneity of soil media was broken and structural heterogeneity of soil media came into the models. This heterogeneity of soil leads to the non-equilibrium flow of water which is commonly known as preferential flow.

Numerous evidence of preferential flow have been observed over the years. Many of the early references to preferential flow primarily consisted of conceptual discussion and qualitative analyses focusing on understanding the nature of the problem rather than quantitative modelling. Lawes et al. (1882) observed that a portion of water escapes by open channels before the body of soil has become saturated. Hursh (1944) while working on forested high infiltration capacity catchments in North Caroline, found that a root-grown channel or wormhole may govern the draining of water through a considerable block of soil. Deeper penetration of infiltrating water closer to tree trunks (Reynolds 1966), E coli movement through forest soil at rates of 15m/h or more (Rahe et al. 1978), and other studies reviewed by Beven (2018) were some of the critical evidence of preferential flows.

Flow through preferred pathways, bypassing most of the soil matrix is known as preferential flow (Jarvis 2007; Nimmo 2020). Various theories and models have been developed for preferential flow processes, but none of them is yet widely accepted. There are three principal modes of preferential flow (Hendrickx and Flury 2001; Nimmo 2020; Franklin et al. 2021). Figure 1 represents the different flow mechanisms along preferential pathways. Of the three main types of preferential flow, macropore flow is most widely studied. Macropore flow is recognized as flow through elongated and larger pores exclusive of the greater bulk of media (Nimmo 2020). Flow processes and modelling of the different phenomenon such as water flow, contaminant transport, and mass transfer between matrix and macropores, depends upon the characteristics of macropores.

2 Characteristics of Macropores

Macropores are open spaces, cavity, passageways, channels, tunnels, or voids in the soil whose characteristic dimensions are significantly larger than the characteristic pore size in the surrounding soil matrix. When water encounters these large voids, it can enter macropores and flow more quickly while avoiding slow movement through the soil matrix (Jarvis et al. 2016). Most macropores form due to old root channels as the root being an organic matter, decays over a period, and leaves void spaces (Aubertin 1971). The physical entrance and expansion of root in the soil compress the matrix and locally changes the porosity of the matrix. As a result, the percentage of pore space thus porosity reduces near the macropores compared to when we move farther from the root channels (Robert and Blevins 1967).

Earthworms and other small burrowing animals also form macropores that are wider than root-grown macropores. Apart from these biotic origins, some non-biotic

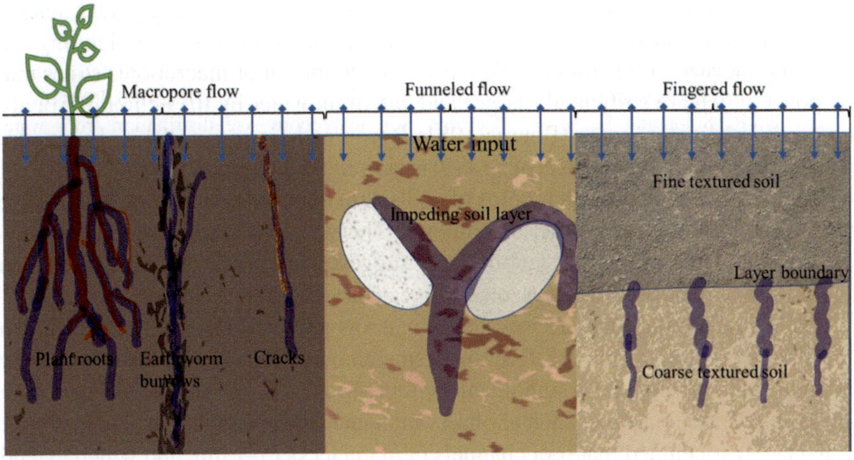

Fig. 1 A schematic representation of different forms of preferential flow. *Source* Franklin et al. (2021). Left portion shows the different form of macropore flow such as due to plant roots growth, earthworm burrows, and shrinkage cracks. The middle portion shows the funnel flow which occurs due to the presence of impeding layers of dense soil media and the right portion represents the fingered flow due to the presence of fine-textured soil above the coarse-textured soil

activities such as freezing and thawing, and shrinkage are also responsible for the formation of soil macropores. Size, shape roughness, and connectivity are the major characteristics of the macropores, which influence the flow processes.

2.1 *Size of Macropores*

The size of macropores can be defined as the aperture size and is quantified in terms of equivalent pore diameter (*EPD*, for circular and irregular pores) (Bouma et al. 1977; Glab 2007) or equivalent width (*EW*, for elongated pores) (Pagliai et al. 1984).

$$EPD = 2\sqrt{\frac{A}{\pi}} \tag{1}$$

$$EW = \frac{-P + \sqrt{P^2 - 16A}}{4} \tag{2}$$

where, A = surface area of the pore, P = perimeter of the pore.

Various techniques such as computer-assisted tomography (Ismail 1975; Perret et al. 1999; Hu et al. 2018), scanning electron microscopy (Jégou et al. 2001), tension infiltrometer (Watson and Luxmoore 1986; Everts et al. 1992), breakthrough curves (Radulovich et al. 1989) have been used to characterize the macropores. Although there is no universal consensus about the demarcation between micro and macropores based on aperture size, researchers have remained interested in defining the characteristic size of macropores. A conceptual definition of macropores states that the pores which are sufficiently larger and can drain water by the action of gravity are considered macropores (Bouma 1981; Beven and Germann 2013). Luxmoore et al. (1990) and Perret Johan et al. (1999) reviewed the literature defining the size of macropores and concluded that pores of equivalent cylindrical diameter in the range of 1–10 mm can be considered macropores. The weight of experimental observations accumulated in the last few decades suggests that pores of EPD larger than 0.3–0.5 mm can be considered as macropores (Jarvis 2007).

It is worth defining characteristic size of macropores by continuity along the depth of soil rather than aperture size. Relatively smaller pores even of size 40 μm or larger (Bouma 1980) can conduct a significant amount of water if continuous throughout the soil profile. Based on experimental observation and fundamental flow theory, Perret et al. (1999) concluded that root channels of diameter 0.75–1 mm and which extend minimum up to 1m in soil can be considered as continous macropores. Earthworm burrows typically 1–3 mm in diameter and that can extend upto 3 to 6 m in vertical length are also categorized as continous macropores (Hendrickx and Flury 2001).

2.2 *Shape of Macropores*

Characterization of the shape of the macropores is essential to understand the flow processes in the macropores. The shape of macropores can be quantified based on the shape index which is defined as the ratio of the area of the pore (A) to the square of its perimeter (P)

$$S_i = \frac{A}{P^2} \tag{3}$$

Primarily, macropores can be classified into 3 broad shape classes, i.e., rounded ($S_i > 0.04$), irregular ($S_i = 0.015$–0.04), and elongated ($S_i < 0.015$) pores (Bouma et al. 1977; Bouma 1980; Pagliai et al. 1983). Rounded or channel-like pores are larger than the pores that could be developed by the packing of soil grains and have a circular cross-sectional shape. Irregular pores or vughs have complex and irregular geometries and are difficult to characterize as they are not inter-connected on vertical or horizontal 2-D thin sections (Bygaart et al. 2000). Elongated or planar pores are sufficiently large with a sheet-like geometry and result due to shrinkage or cracks mostly in clayey soils.

Electro-optical analysis and micro-morphological studies show that in a typical soil, 50–60% of pores are rounded, 30–40% of pores are irregular and 10–12% of pores are elongated. However, elongated pores are larger and extend deeper in the soil thus contributing a major fraction (around 70–80%) of soil macroporosity. Again, there is no quantitative relation between the shape of macropores and the type of soil, however, the mode of formation influences the shape of macropores. As an example, rounded pores are generated due to biological activities and distributed in the upper layer of soil, whereas shrinkage of drying soil creates planar voids.

2.3 Roughness of Macropore Wall

The surface roughness of macropore walls plays a significant role in different flow and transport processes. It affects the flux rate, the residence time of water and solutes within the macropores, and flow patterns. These surface characteristics govern the mass transfer between the macropores and the diffuse domain. The exchange term between domains can be accurately modeled if the physio-chemical properties of macropore surfaces are known to us (Gerke 2012). Generally Scanning Electron Microscopy (SEM) techniques cannot be used to measure the elevation of different surface points and Confocal Laser Scanning Microscope (CLSM) is used for measuring the elevation. The structural roughness of macropore can be quantified as the root mean square elevation of surface points (Leue and Gerke 2016).

$$R_q = \sqrt{\frac{1}{N}\sum_{n=1}^{N} Z_n^2} \tag{4}$$

where N is the total number of measurement points and Z_n is the elevation of nth measurements.

Studies on the roughness characterization show that the roughness of the wall is governed by the formation of macropores and the type of soil (Leue and Gerke 2016). In fine-textured soil, illuviation processes reduce the surface roughness. The structural roughness of the macropore wall is lesser if the pores are generated due to root growth processes than if they are generated by earthworms as roots apply higher pressure on the wall which changes the arrangement of soil particles along the pore wall. Lesser roughness leads to a higher velocity of flow and less microbial activity along the wall. Generally, the roughness of root-grown macropore walls ranges between 10 and20 μm whereas 20–40 μm in the case of earthworm burrows.

2.4 Functional Macropores

The above characteristics cannot completely provide the flow information through the macropores. Therefore, an important aspect related to the functionality of the macropores is that the pores should be continuous and vertically connected to facilitate water and air transport throughout the soil profile (Budhathoki et al. 2022). Not all elongated structural voids can be activated during flow processes. An active macropore network is a subset of a total network that can be quantified by using dye tracer studies in addition to image analysis (Mooney and Morris 2008). Sammartino et al. (2015) proposed a new method of image acquisition based on recording water infiltration by a sequence of 3-D images to characterize the active macropore network. Poor interconnectivity (Warner et al. 1989), dead ends (Urbina et al. 2019) in the soil matrix, and entrapped air (Luo et al. 2008) lead to the inactiveness of macropores (Table 1).

3 Macropore Flow in the Field Soil

As soon as a land surface gets input of water, some fraction of it goes into the subsurface through macropores. It is important to note that a major fraction of flow travels a significant distance through macropores which constitutes only a smaller fraction of the medium's volume. In most of the experiments, macroporosity of soil is found between 5 and 15% (HU et al. 2018; Pagliai et al. 1984; Watson and Luxmoore 1986) whereas it constitutes 65–80% of the flow. Figure 2 depicts the macropore flow processes in the field. Some of the macropores which start from the land surface get water directly ($q(0, t)$) from water sources ($q^*(t)$) or overland flow run on ($o(t)$). Below the soil surface, seepage from the soil matrix can produce a significant amount of flow through the macropores. As macropore flow occurs generally at a faster velocity, it would be in non-equilibrium with the surrounding soil matrix; however, a certain portion of water transfers from the macropore to the matrix ($r.\ w(z, t)$).

3.1 *Forms of Macropore Flow*

Flow in unsaturated macropores can occur either in the form of capillary flow or film flow (Dragila and Wheatcraft 2001). Experimental observations provide some evidence of the intermittent flow in the macropores. As in capillary flow, water first enters smaller interstices, and with increasingly wet conditions (matric potential approaches to zero), water occupies macropores (Miller and Miller 1955; Peters and Klavetter 1988). An important aspect of capillary flow is that fluid will be in contact of the macropore wall in all directions. A full continuous saturated pathway is required to conduct the flow in the soil media. As per this theory, at lower capillary potential (unsaturated soil), the largest water filled macropore will be of the same order of largest water filled pores of surrounded soil matrix in order to maintain the local equilibrium.

Capillary flow theory for macropore flow cannot explain the anomalous high-speed flow (Beven and Germann 1982; Nimmo 2007; Tokunaga and Wan 1997) and transport through macropores. A free surface film flow mechanism by which water in the unsaturated macropores may move with higher velocity than predicted by capillary flow theory can provide a better understanding of the recently observed

Table 1 Studies on macropore characterization

Articles	Topics covered	Major highlights
Aubertin (1971)	Soil macropores in forest soil	Biotic sources of macropore formations: root channels, earthworm burrows
Ismail (1975)	Micro-morphometric soil porosity characterization	Estimation of different void patterns using Electro-optical image analysis
Bouma et al. (1977)	Functions of different types of macropore during saturated flow	Different shape classes of macropores were observed with their size distribution
(Bouma (1980)	Soil morphology and preferential flow	A distinction is made between the physical, morphological, and dye tracer studies for characterizing soil macropores
Luxmoore (1981)	Micro, meso, and macroporosity of soil	Suggests the realistic size of macropore and threshold water entry pressure
Beven and Germann (1982)	Macropore flow in field soil	Reviews the macropore definition and its size classification while addressing the selection of representative elementary volume (REV) for the continuum approach in macroporous soil
Pagliai et al. (1983)	Micro-morphometric and micromorphological investigation of soil	Estimates macroporosity in No-tilled and conventional tilled plot. Conventional tilled plot reflects higher value of macroporosity with lesser number of macropores
Bronswijk (1988)	Soil macropores in clayey soil	Models bypass flow by quantifying the shrinkage generated macropores in clayey soil
Luxmoore et al. (1990)	Physical characteristics of macropores	Reviews the porosity terminologies such as micro, meso, and macro pores
Perret et al. (1999)	3-D quantification of macropores in undisturbed soil cores	Discusses different characteristics of macropores such as shape, size, tortuosity, coordination number, numerical density, and connectivity using X-Ray computer-assisted tomography
Glab (2007)	Image analysis for macropore characterization	Assesses the presence of macropores based on their actual pore diameter for quantifying the soil macroporosity
Luo et al. (2008)	Quantification of soil structure using X-Ray computer tomography and breakthrough curve	Quantifies the variation of properties of macropores along the depth of soil profile and their ability to exchange mass with soil matrix
Josa et al. (2013)	Influence of tillage practices on soil macropore distribution	Show that the presence of larger macropores is higher in conventional tillage than that in no-tillage
Leue and Gerke (2016)	Roughness of bio-pores and cracks	Highlights the importance of roughness of macropore walls and its measurement using CLSM
Leue et al. (2019)	Separation of soil macropore type based on 3-D Tomography	Estimates the different shape and size parameters for bio pores and coated cracks

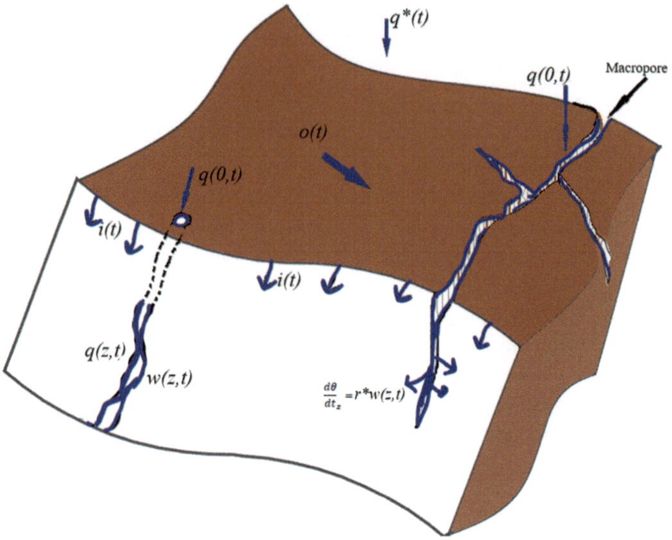

Fig. 2 Macropore flow in the field soil. *Source* Beven and Germann (1982)

behavior of macropore flow and transport (Tokunaga and Wan 1997; Dragila and Wheatcraft 2001). The film flow model consists of water flowing down the wall of air-filled macropores in which the water film is in contact with one of the walls of macropores and sustains an air phase between the film and the opposing face wall of macropores (Fig. 3). The film flow mechanism is governed by gravity and viscous forces with very little inclusion of capillary flow (Nimmo 2020). Free surface film theories differ from conventional macropore flow theories in that the size of the macropore does not control the flow rate. Film flow can occur in small and large apertures at the same time and at similar moisture states of the surrounding soil matrix.

Some laboratory experiments (Su et al. 1999) show the intermittent behavior of macropore flow. Based on the prevailing conditions (such as water source or moisture state) and macropore size distribution (fluid flowing through large to small to large aperture), macropore flow can either actively conduct the water flow or can be temporarily inactive. Overall, there is a diversity of opinions concerning which form of the flow governs the macropore flow. Nevertheless, the major challenge associated with all these theories is how to include these pore or profile scale mechanisms in field-scale models.

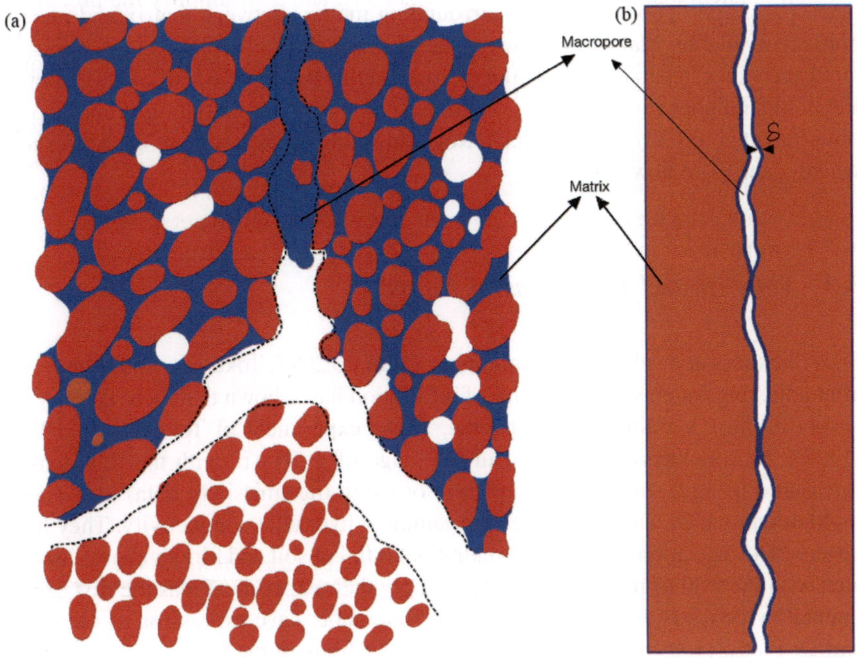

Fig. 3 A schematic representation of different forms of macropore flow. *Source* Dragila and Wheatcraft (2001)

3.2 Initiation of Macropore Flow

Predictive macropore flow models employ different criteria for the initiation of macropore flow. The time lag between input and initiation of macropore flow can be a few minutes to several hours depending on the threshold values of (1) input intensities such as precipitation, run-on, or ponding; (2) moisture content of the surrounding soil matrix; (3) matric potential typically equal to water entry pressure of minimum size of macropore and (4) infiltration capacity of surrounded soil. Capillary flow theory states that macropore flow occurs when water pressure locally increases to near saturation at the interface of the matrix and macropore. An experimental study by Cey et al. (2009) shows that a certain threshold value of input flux (12–20 mm/h) or pressure head (– 3 cm of water) is required to initiate the macropore flow. Matrix infiltration capacity is also a strong indicator of the macropore flow initiation and behavior (Nimmo 2016).

Experimental evidence show that macropore flow can start without meeting any of the above criteria. Nimmo (2012) has cited a few experimental studies which clearly signify that even the drier matrix can lead to the macropore flow. As macropore flow occurs under the condition of hydraulic disequilibrium, it initiates if the contrasting conditions (such as contrast between pore sizes, moisture state between matrix and macropore, hydrophobicity of the surrounding matrix) between matrix and macropore are available. Macropore flow occurs through the soil which is partially saturated and water bypasses most of the matrix and flows along the wall of macropores. Reynold's number can also be a good representation of the initiation of macropore flow. Tao et al. (2020) reported that a Reynolds number greater than 3 is a threshold where macropore flow begins.

3.3 *Velocity and Thickness of Films*

Film thickness and velocity measurements (Nimmo 2007; Tokunaga and Wan 1997) after adjusting for episodic behavior of infiltration have shown that there is less variability in these variables than expected. In the experiment of Tokunaga and Wan (1997), film thickness is observed in the range of 2–70 μm with the flux density (measured using a smooth flow equation or using transmissivity data) of 2 m/day to 40 m/day which was 1000 times the saturated hydraulic conductivity. They have observed a correlation between the matric potential of soil and film thickness associated with the wall of macropores. Interestingly their experiment was the first experimental evidence of film flow in the soil. Kinematic wave expression of Germann and Di Pietro (1999) computes the thickness of films in the range of 0.31–65 μm which is comparable with the results of Germann and Hensel (2006). Germann et al. (2007) reported the film thickness to be in the range of 2.4–11.5 μm which is maintained over a depth of 10–20 cm. Using water content wave approach, Hincapié and Germann (2009a, b) estimated the film thickness and noted that it varies in the range of 6–120 μm. Based on above observations we can conclude that the thickness of film flow varies in the range of 3–100 μm and depends upon the surface characteristics of the macropore and the properties of water. Again, in most of the soil media, these properties vary less than the variation in geometrical size of the pore. Thus, the thickness of film flow remains more or less constant during any macropore flow event.

The non-equilibrium nature of macropore flow creates a lower limit for film velocity. If the flow velocity is less than that, the flow will be absorbed by the matrix before it travels to a larger depth in the soil. Field observed macropore flow occurs at a velocity speed in the range of 10^{-4}–3×10^{-3} m/s for various geometrical and hydraulic properties of soil media. Using meta-analysis over the 243 observations cited in 76 articles, (Gao et al. (2018) showed that macropore flow velocity has a geometrical mean of 1.08×10^{-3} m/s. Macropore flow velocity has a strong correlation with the supply of water and thickness supported by the wall. Moreover, evidence suggests that flow velocity through macropores at field scale is greatly affected by the size of pores.

3.4 Current Understanding of Macropore Flow Processes

The acknowledgement of macropore flow as a legitimate phenomenon in heterogenous soil media only occurred after 1984, hindering to a considerable extent our understanding of the processes of macropore flow. The complexity of factors such as antecedent moisture, pore geometry, mass transfer between matrix and macropore, and the characteristics of input sources make it more difficult to understand macropore flow. As water flows in soil pores of all sizes under the influence of capillarity, gravity, and viscous forces, in larger pores, capillarity is not so large making gravity the dominant force for conveying water through it. Here surface tension in terms of capillary flow cannot control the flow in individual pores like in traditional theories of porous media but plays a major role in conjunction with gravity and viscosity to determine the flow-controlling geometrical configuration of film in macropores.

There is no consensus theory available for the initiation of macropore flow; however, recent observations refer that water can preferentially flow through larger pores without saturating the porous media. Velocity observed in natural macropore flow conditions is lower than that predicted using the Poiseuille equation. Some numerical studies based on the field-observed flux values for macropore flow have been carried out to estimate the radius of filled pores using the Poiseuille equation (Germann and Hensel 2006). The computed radii often range from 10 to 30 μm. Macropores of these aperture sizes at a length scale of several cm to m are not naturally possible in the soil. These studies refer to the fact that macropore flow can take place even though these pores are not entirely filled with water. The flow capacity of these macropores is sufficiently larger than the water supply so the flow conditions remain unsaturated in the macropores (Jarvis et al. 2016).

4 Modelling of Macropore Flow

Macropore flow results in irregular wetting of the soil profile unlike uniform wetting (Green and Ampt 1911; Richards 1931) of homogeneous soil. Thus, an important aspect of macropore flow is that a part of the wetting front, while bypassing most of the matrix, can advance more quickly than the modelled wetting front obtained using Richards equation. Even in the case of uniform flow, macropore flow occurs contrary to Richard's assumptions of equilibrium flow. For modelling flow through heterogeneous pore structures, earlier researchers used the Richards equation with certain modifications in the flow system like a dual porosity system (Durner 1994) with the van Genuchten–Mualem (van Genuchten 1980) model to describe the soil water characteristic curve for each domain. These models predict the rapid increase in hydraulic conductivity near saturation (Mohanty et al. 1997) although they are based on the equilibrium assumption.

Over the past four decades, researchers have employed various strategies to account for the characteristics of macropore flow in macropore flow models. Most of the existing models take a macroscopic continuum approach by treating macropores as a separate domain in the soil system, characterized by distinct hydraulic properties that differ from the surrounding soil matrix. These different approaches to model the macropore flow have been examined in the subsequent sections.

4.1 Richards-Based Approach

Classical Richards equation is the common approach to model the fluid flow produced by gravity and pressure gradient force in the unsaturated porous media. This equation assumes local equilibrium between moisture content and pressure head at each point in the porous media. The relationship between moisture content and pressure head is typically represented in the form of soil water retention characteristic functions (Brooks and Corey 1966; van Genuchten 1980). Flow through soil macropores can also be modelled using the Richards equation by using a dual-domain approach to account for the non-equilibrium nature of macropore flow. In the dual-domain approach, soil matrix and macropores are treated as separate domains and in both domains, water flow is controlled by capillary and gravity forces. While modelling the macropore domain, the macropore network can be replaced by a porous media with different hydraulic properties (see Fig. 4).

Single porosity: Ross and Smettem (2000) proposed a simplistic model to treat the non-equilibrium of flow by removing the Richards assumption of coupling relation

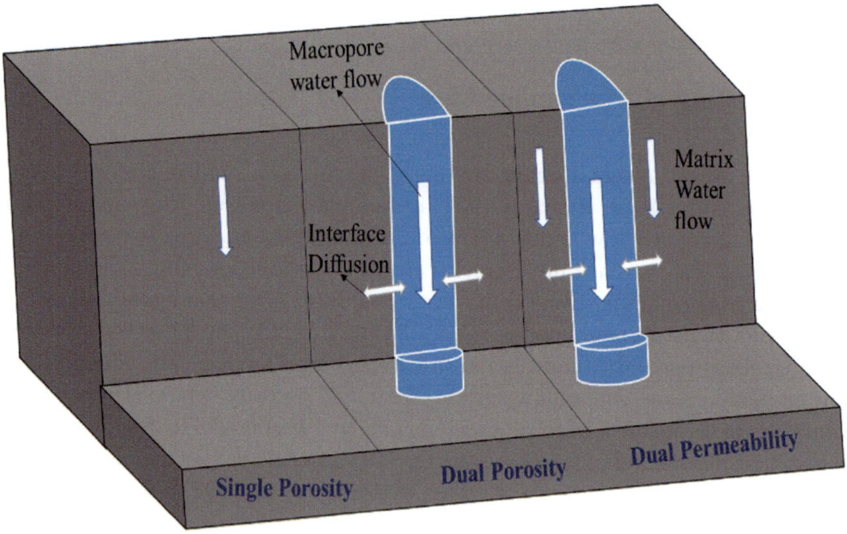

Fig. 4 Classification of conceptual models of water movement and contaminant transport in Macroporous soil media. *Source* Orozco-López et al. (2018)

between moisture content (θ) and matric potential (h). This hypothesis leads to the two independent variables (θ and h). By combining Richards equation (Eq. 5) with the kinetic water content equation (Eq. 6), a coupled system of equations is obtained which can be solved using any numerical scheme.

$$\frac{\partial \theta}{\partial t} = \frac{\partial}{\partial x}\left(K(h)\left(\frac{\partial h}{\partial z} + 1\right)\right) \tag{5}$$

$$\frac{\partial \theta}{\partial t} = f(\theta, \theta e) = \frac{(\theta e - \theta)}{\tau} \tag{6}$$

where θ is actual water content, *K(h)* is unsaturated hydraulic conductivity, θe is the equilibrium moisture content, and τ is equilibration time constant which describes the rate at which the soil water content responds to changes in the pressure head at the soil surface. A higher value of τ produces faster moving wetting front which can travel to a greater depth than those produced with other models. This formulation offers the advantage of incorporating the macropore flow using only one additional parameter (τ). Estimation of hydraulic conductivity as a function of pressure head rather than as a function of moisture content, reflects the nonequilibrium nature of the flow system in which flow occurs contrary to Richards' assumption. The physical description of macropore geometry is not considered while modelling the flow using a single porosity approach. Šimůnek et al. (2003) reported numerical instabilities in the models while using larger value of equilibration time constant.

Dual-porosity modelling. In the dual porosity model (Genuchten and Wierenga 1976), the domain representing the soil matrix is typically considered as a stagnant or immobile zone which can exchange and store water, but the convective flow of water is not allowed. Water movement in soil matrix primarily occurs due to the diffusion process. In contrast to the single porosity model, this model partitions the moisture content into mobile and immobile region (Eq. 7). The water dynamics can be represented using the Richards equation in the mobile region (Eq. 8) whereas a mass balance equation to describe the moisture variation in immobile domain (Eq. 9).

$$\theta = \theta_{im} + \theta_m \tag{7}$$

$$\frac{\partial \theta_m}{\partial t} = \frac{\partial}{\partial z} K(h_m)\left(\frac{\partial h}{\partial z} + 1\right) - S_m - \Gamma_w \tag{8}$$

$$\frac{\partial \theta_{im}}{\partial t} = -S_{im} + \Gamma_w \tag{9}$$

where subscript *m* is for mobile and subscript *im* is for an immobile domain, S_m and S_{im} are source/sink term for mobile and immobile domains, respectively, Γ_w represents mass transfer from macropore to matrix, which is proportional to the difference of effective water content (θe) of two domains (Šimůnek et al. 2001), and α is first order rate coefficient.

$$\Gamma_w = \alpha\left(\theta_e^m - \theta_e^{im}\right) \tag{10}$$

It is worth noting that dual porosity formulation is commonly used for solute and contaminant transport studies. When it comes to water movement in heterogeneous porous media, dual porosity models work at low flux rate only. This is because flow modelling for higher fluxes often requires a more detailed description of porous media which is not considered in dual porosity formulation.

Dual permeability: In the dual permeability models (Gerke and van Genuchten 1993a), water flow in both the macropore and the matrix domain is assumed as Darcy-Richards' flow. The matrix domain represents a low permeability domain. On the other hand, a macropore domain represents high permeability pathways. A separate Richards' equation can be employed for both domains with different sets of parameters.

$$\frac{\partial \theta_f}{\partial t} = \frac{\partial}{\partial z}\left(K_f(h_f)\left(\frac{\partial h_f}{\partial z} + 1\right)\right) - S_f - \frac{\Gamma_w}{w_f} \tag{11}$$

$$\frac{\partial \theta_m}{\partial t} = \frac{\partial}{\partial z}\left(K_m(h_m)\left(\frac{\partial h_m}{\partial z} + 1\right)\right) - S_m + \frac{\Gamma_w}{1 - w_f} \tag{12}$$

where subscript f represents macropore domain, subscript m represents matrix domain. K_f is macropore hydraulic conductivity as a function of macropore pressure head (h_f), K_m is matrix conductivity which is function of matrix pressure head (h_m), S represents source/sink term, Γ_w is the rate of water exchange between matrix and macropore, and w_f is dimensionless volume factor defined as ratio of macropore domain volume to total soil volume. First order mass transfer function is given by:

$$\Gamma_w = \alpha\left(h_f - h_m\right) \tag{13}$$

in which α is first order mass transfer coefficient which can be estimated as

$$\alpha = \frac{\beta}{a^2} K_a \gamma_w. \tag{14}$$

where β is shape factor, a is characteristic length of aggregate, γ_w is scaling factor and K_a is the hydraulic conductivity of matrix-macropore interface. Mass transfer can be represented in higher orders terms (Arora et al. 2011; Gerke and van Genuchten 1993b)

Multi porosity and multi permeability model: Multi-porosity and multi-permeability models have more than 2 domains (such as macropore, mesopore and micropore (Gwo et al. 1995)). Flow processes of each domain are assumed as Darcy-Richards' flow, and Richards equation can be applied for all the domain with different set of parameter for each domain. Although these types of formulation provide much information about the porous media, they are computationally expensive and require many parameters.

Limitations of Richards approach. A significant drawback of Richards-based approach is their substantial parameter requirement for accurately representing the matrix and macropore domains. The presence of large number of parameters can make the model calibration process challenging, as it becomes more difficult to uniquely estimate the optimal value of parameters. Again, a many parameters model is flexible to fit diverse datasets, solely due to its abundant degree of freedom which does not necessarily align with the underlying physics. In its full complexity, dual permeability model requires 16 parameters, six parameters each to characterize the soil matrix ($\theta_s^m, \theta_r^m, \alpha_m, n_m, l_m, K_s^m$) and macropores ($\theta_s^f, \theta_r^f, \alpha_f, n_f, l_f, K_s^f$) and four parameters to define interface mass transfer (*w*, K_{as}, α_a, n_a). Total 11 parameters are required to model the flow using dual porosity based on the Richards equation with van Genuchten–Mualem model for soil hydraulic function and pressure head based mass transfer term (six parameters for macropore ($\theta_s^f, \theta_r^f, \alpha_f, n_f, l_f, K_s^f$), , four parameters for matrix ($\theta_s^m, \theta_r^m, \alpha_m, n_m$) and one parameter for interface mass transfer (K_{as})).

Richards' equation is based on the experiments performed under the controlled lab conditions in homogeneous soil in which flow occurs under the action of capillarity and gravity forces. In the case of macropore flow, flow is predominantly controlled by gravity and viscous forces. Also, Richards' based models cannot capture the effect of variation of water source at surface to a greater depth as these models assume vertical homogeneity in the media while it is not actually true in natural soil media (Orozco-López et al. 2018). So, if Richards based approaches are inadequate to model the macropore flow, is there an alternative approach that could be more suitable?

4.2 *Stokes Flow Approach*

Stokes flow provides a quantitative framework for modelling the macropore flow where flow occurs at a low Reynolds number. Stokes flow can be applied in such cases where gravity acts as a driving force and viscosity helps in dissipating momentum within water films moving along the solid wall with a free surface. The linear momentum equation for modest tortuous media is given by:

$$\theta \frac{\partial (q_f/\theta)}{\partial t} = -g\nabla(\psi + z) + \eta\nabla^2 q_f + \eta B q_f \tag{15}$$

here θ is moisture content, q_f is volume flux density in fracture (m/sec), ψ is matric potential, η is kinematic viscosity of water, *g* is the acceleration due to gravity, *B* is second order tensor related to liquid solid contact area. With suitable boundary conditions and assumptions, Germann and Di Pietro (1999) have derived a simple equation for volume flux density of mobile water which is given by:

$$q_f(F, L) = \frac{g}{3\eta} L F^3 \tag{16}$$

where F is maximum film thickness and L is the effective contact length per unit cross-sectional area. Several models are available which are based on the Stokes flow assumption for macropore flow which can be broadly classified in two categories.

Kinematic wave modelling: Kinematic wave theory (Germann and Beven 1985; Hincapié and Germann 2009a, b) is the common approach to model the macropore flow. This model provides an advantage over dual porosity type models in which vertical heterogeneity of the media were neglected. Kinematic wave equation describes infiltration and redistribution of single rectangular pulse of water in the macropore. Kinematic wave based dual porosity models assume that flow in the soil matrix occurs under the action of capillarity and gravity while gravity itself drives the macropore flow. The mass balance equation in macropore domain was given as described by Germann (1985) and modified by Germann and Beven (1985) by adding sink term.

$$\frac{\partial q_f}{\partial \mathrm{t}} + c\frac{\partial q_f}{\partial \mathrm{z}} + c\theta\alpha = 0 \tag{17}$$

where q_f is volumetric flux density which is given as $q_f = a\ \theta_f^b$ (Lighthill and Whitham 1955), a is macropore conductance, b is kinematic exponent, c is celerity defined as change in volumetric flux with respect to change in mobile water content (Germann and Di Pietro 1999).

$$c = \frac{\partial q_f}{\partial \theta_f} = \frac{g}{\eta} F^2 \tag{18}$$

Mass exchange rate α is given as

$$\alpha = -\frac{1}{\theta_f}\frac{\partial \theta_f}{\partial \mathrm{t}} \tag{19}$$

The kinematic wave based dual porosity model requires a total of 10 parameters (1 less than that required by conventional dual porosity model). Matrix requires six parameters ($\theta_s^m, \theta_r^m, \alpha_m, n_m, l_m, K_s^m$) while macropore and interface require four parameters (θ_f, a, b, d where$d$ is diffusion path length). The kinematic wave model can predict vertical downward infiltration, but it cannot model upward evaporation as this model is limited to gravity driven flow only. MACRO model (Jarvis et al. 1994) which is a dual permeability type model uses power law function (Eq. 20) to simulate flow in macropore domain which reduces the number of parameters significantly.

$$q = a\left(\frac{\theta_f}{\theta_{sf}}\right)^b \tag{20}$$

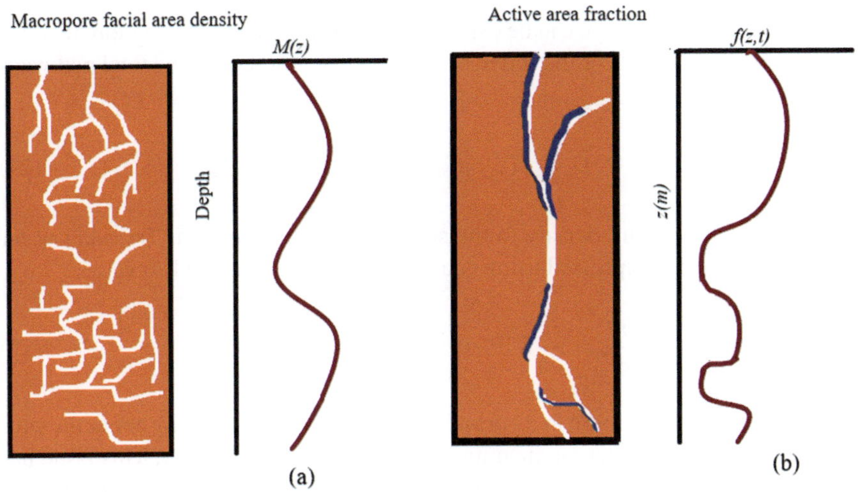

Fig. 5 Physical definition of model parameter. *Source* Nimmo and Mitchell (2013)

where θ_{sf} is saturated water content of the macropore.

Source responsive modelling: Source-responsive modelling (Nimmo 2010; Nimmo and Mitchell 2013) approach is similar to dual permeability-based model with diffuse and source responsive domain. Diffuse domains are always active although with slow dynamics, while source responsive domain is sometimes active. Source responsive domain is characterized using two parameters i.e., macropore facial area density *(M)* and active area fraction (f). *M* represents capacity for macropore flow while f represents its degree of activation. Figure 5 represents these parameters in macroporous soils.

Flux and moisture content can be partitioned like other dual domain models and can be given using the following expressions

$$\theta = \theta_D + \theta_s \tag{21}$$

$$q = q_D + q_s \tag{22}$$

where, subscript D is for diffuse domain and subscript S is for source responsive domain, θ_D is volumetric moisture content of diffuse domain, θ_s is volumetric moisture content of source responsive domain, q_D is flux through diffuse domain, q_s is flux through source responsive domain, θ is total soil water content and q is total flux. Flux through diffuse domain is expressed by Darcy-Buckingham equation as below.

$$q_D(z,t) = K(\theta_D)\frac{\partial(\psi + z)}{\partial z} \tag{23}$$

Flux through macropore can be expressed by Eq. 16 with multiplication of active area fraction to account for the active network of macropore over the total pathways available.

$$q_S(z,t) = f(z,t)q_{S\max} = \frac{1}{3}\frac{g}{\nu}F^3 M f(z,t) \tag{24}$$

where $q_{s\max}$ is max flux density in the source responsive domain. Transient water content dynamics is expressed using continuity equation as given below.

$$\frac{\partial \theta}{\partial t} = \frac{\partial}{\partial z}K(\theta_D)\left(\frac{\partial \psi}{\partial z} + 1\right) - \frac{\partial f q_{s\max}}{\partial z} \tag{25}$$

Source-responsive model in its primary stage had some ambiguity about the suitable REV for which different quantities can be added algebraically. This issue has been partially resolved by Masciopinto (2012) by averaging diffuse domain over the source responsive domain REVs as following.

$$\theta = \theta_s + \int_{REV_S} \theta_D dV \tag{26}$$

Similar considerations must be made for flux density and other variables.

Limitations of Stokes-based approach. Adoption of Stokes flow assumption in macropore flow modelling closely reflects the physics of flow. Nevertheless, a compromise can be found between complexity and parameter identifiability at different application scales (Guertault and Fox 2020). These models which have been developed based on pore scale mechanisms do not address the change of physics when scale of study (pore scale to field scale) changes.

4.3 Needs for Improvement in Modelling Practices

The scientific community has made significant progress in understanding macropore flow processes (Jarvis et al. 2016) which is also reflected in the diverse modelling approaches used by different researchers. However, the models developed so far need to be examined for their physics. For instance, a widely utilized approach is Stokes based modelling for gravity driven film along macropore walls but it does not include the flow path geometries (shape and roughness of macropores). The connectivity of the macropore should also be characterized appropriately in the numerical models. As macropore flow is highly intermittent and not a continuous flow process, the flow models should be sensitive to initial and boundary conditions. Although source-responsive model (Nimmo 2010) is sensitive to input intensities at the surface of soil and is based on the Stokes flow assumption, research is required to adequately define the model parameters which are empirical in nature.

One of the major issues in macropore flow modelling is selecting the scale for different domains in dual domain modelling approaches. For example: representative elementary volume (REV) for homogeneous soil is 10^{-3} m^3 or more, whereas macropore flow pathways can extend up to 1 m and require a larger REV. Discrete macropore pathways with continuum matrix characterization can assist in addressing scale issues for macropore flow modelling. Another problem is related to the mass transfer between the domains. Macropore flow can start at land surface and can get fed at different depths by the soil matrix. Which property of soil matrix is required to characterize the transfer term is a topic of active research. Understanding the major cause considering wetness of matrix, nature and duration of water input, types of macropore can help in solving these issues.

5 Concluding Remarks

Macropores are the larger pores in soil in which flow is majorly governed by the gravity and viscous forces. Flow occurs through these macropores while bypassing the soil matrix. This hampers the accurate prediction of flow and transport in heterogenous soil media and can significantly affect the estimation of crop water requirement, contaminant and fertilizer transport, ground water table variation etc. Again, for such an important phenomenon, the scientific community has not yet settled on any approach for modelling. Selection of an appropriate model involves a reasonable balance between physical realism and usability considering the specific purpose, available computational power and constraints of the study or application. While film flow theories based on Stokes assumption can predict the macropore flow accurately with lesser number of parameters, they do not conceptualize the geometries of the macropores well. On the other hand, Richards based dual domain models can predict macropore flow, but they require a large set of parameters. Also, research is needed to comprehensively understand the interdomain mass transfer and mechanisms responsible for it in macroporous soils.

References

Arora B, Mohanty BP, McGuire JT (2011) Inverse estimation of parameters for multidomain flow models in soil columns with different macropore densities. Water Resour Res 47(4). https://doi.org/10.1029/2010WR009451

Aubertin GM (1971) Nature and extent of macropores in forest soils and their influence on subsurface water movement (vol 192). Northeastern Forest Experiment Station

Beven K (2018) A century of denial: preferential and nonequilibrium water flow in soils, 1864–1984. Vadose Zone J 17(1):1–17. https://doi.org/10.2136/VZJ2018.08.0153

Beven K, Germann P (1982) Macropores and water flow in soils. Water Resour Res 18(5):1311–1325. https://doi.org/10.1029/WR018I005P01311

Beven K, Germann P (2013) Macropores and water flow in soils revisited. Water Resour Res 49(6):3071–3092. https://doi.org/10.1002/WRCR.20156
Blevins RL (1967) Micromorphology of soil fabric at tree root-soil interface. The Ohio State University
Bouma J (1981) Soil morphology and preferential flow along macropores. Agric Water Manag 3(4):235–250. https://doi.org/10.1016/0378-3774(81)90009-3
Bouma J, Jongerius A, Boersma O, Jager A, Schoonderbeek D (1977) The function of different types of macropores during saturated flow through four swelling soil horizons 1
Bouma J (1980) Soil Morphology and preferential flow along macro-pores I. In: Agricultural water management, vol 3
Brooks RH, Corey AT (1966) Properties of porous media affecting fluid flow. J Irrig Drain Div 92(2):61–88. https://doi.org/10.1061/JRCEA4.0000425
Bronswijk JJB (1988) Modeling of water balance, cracking and subsidence of clay soils. J Hydrol 97(3–4):199–212
Budhathoki S, Lamba J, Srivastava P, Malhotra K, Way TR, Katuwal S (2022) Using X-ray computed tomography to quantify variability in soil macropore characteristics in pastures. Soil Tillage Res 215:105194. https://doi.org/10.1016/J.STILL.2021.105194
Buckingham E (1907) Studies on the movement of soil moisture
Cey EE, Rudolph DL, Passmore J (2009) Influence of macroporosity on preferential solute and colloid transport in unsaturated field soils. J Contam Hydrol 107(1–2):45–57. https://doi.org/10.1016/J.JCONHYD.2009.03.004
Dragila MI, Wheatcraft SW (2001) Free-surface films. In: Conceptual models of flow and transport in the fractured vadose zone
Durner W (1994) Hydraulic conductivity estimation for soils with heterogeneous pore structure. Water Resour Res 30(2):211–223
Everts CJ, Kanwar RS, Asae A (1992) Interpreting tension-infiltrometer data for quantifying soil macropores: some practical considerations
Franklin SM, Kravchenko AN, Vargas R, Vasilas B, Fuhrmann JJ, Jin Y (2021) The unexplored role of preferential flow in soil carbon dynamics. Soil Biol Biochem 161:108398. https://doi.org/10.1016/J.SOILBIO.2021.108398
Gao M, Li HY, Liu D, Tang J, Chen X, Chen X, Blöschl G, Ruby Leung L (2018) Identifying the dominant controls on macropore flow velocity in soils: a meta-analysis. J Hydrol 567:590–604. https://doi.org/10.1016/J.JHYDROL.2018.10.044
van Genuchten MT, Wierenga PJ (1976) Mass transfer studies in Sorbing porous media I. analytical solutions. Soil Sci Soc Am J 40(4):473–480. https://doi.org/10.2136/SSSAJ1976.03615995004000040011X
Gerke HH, van Genuchten MT (1993a) A dual-porosity model for simulating the preferential movement of water and solutes in structured porous media. Water Resour Res 29(2):305–319. https://doi.org/10.1029/92WR02339
Gerke HH, van Genuchten MT (1993b) Evaluation of a first-order water transfer term for variably saturated dual-porosity flow models. Water Resour Res 29(4):1225–1238. https://doi.org/10.1029/92WR02467
Gerke HH (2012) Macroscopic representation of the interface between flow domains in structured soil. Vadose Zone J 11(3):vzj2011.0125. https://doi.org/10.2136/VZJ2011.0125
Germann PF, Beven K (1985) Kinematic wave approximation to infiltration into soils with sorbing macropores. Water Resour Res 21(7):990–996. https://doi.org/10.1029/WR021I007P00990
Germann PF, Di Pietro L (1999) Scales and dimensions of momentum dissipation during preferential flow in soils. Water Resour Res 35(5):1443–1454. https://doi.org/10.1029/1998WR900112
Germann PF, Hensel D (2006) Poiseuille flow geometry inferred from velocities of wetting fronts in soils. Vadose Zone J 5(3):867–876. https://doi.org/10.2136/VZJ2005.0080
Germann P, Helbling A, Vadilonga T (2007) Rivulet approach to rates of preferential infiltration. Vadose Zone J 6(2):207–220. https://doi.org/10.2136/VZJ2006.0115

Germann PF (1985) Kinematic wave approach to infiltration and drainage into and from soil macropores. Trans ASAE 28(3):745–0749. https://doi.org/10.13031/2013.32331
Glab T (2007) Application_of_image_analysis_for_s
Guertault L, Fox GA (2020) Performance of preferential flow models in predicting infiltration through a remolded soil with artificial macropores. Vadose Zone J 19(1):e20055. https://doi.org/10.1002/VZJ2.20055
Gwo JP, Jardine PM, Wilson GV, Yeh GT (1995) A multiple-pore-region concept to modeling mass transfer in subsurface media. J Hydrol 164(1–4):217–237. https://doi.org/10.1016/0022-1694(94)02555-P
Haines WB (1930) Studies in the physical properties of soil. V. The hysteresis effect in capillary properties, and the modes of moisture distribution associated therewith. J Agr Sci 20(1):97–116. https://doi.org/10.1017/S002185960008864X
Heber Green W, Ampt GA (1911) Studies on soil phyics. J Agric Sci 4(1):1–24. https://doi.org/10.1017/S0021859600001441
Hendrickx JMH, Flury M (2001) Uniform and preferential flow mechanisms in the vadose zone. https://www.researchgate.net/publication/318458237
Hincapié IA, Germann PF (2009a) Abstraction from infiltrating water content waves during weak viscous flows. Vadose Zone J 8(4):996–1003. https://doi.org/10.2136/VZJ2009.0012
Hincapié I, Germann P (2009b) Gravity-Driven viscous flow in sand boxes assessed with neutron radiography. Vadose Zone J 8(4):891–901. https://doi.org/10.2136/VZJ2009.0003
Hu X, Li Z, Li X, Wang P, Zhao Y, Liu L, Lü Y (2018) Soil Macropore structure characterized by x-ray computed tomography under different land uses in the Qinghai lake watershed Qinghai-Tibet Plateau. Pedosphere 28(3):478–487. https://doi.org/10.1016/S1002-0160(17)60334-5
Hursh CR (1944) Appendix B—Report of sub-committee on subsurface-flow. Eos Trans Am Geophys Union 25(5):743–746
In Forest Soils and their influence on subsurface water movement forest service (1971)
Ismail SNA (1975) Soil Survey Papers, No. 9 Micromorphometric soil-porosity characterization by means of electro-optical image analysis (Quantimet 720)
Jarvis NJ (2007) A review of non-equilibrium water flow and solute transport in soil macropores: principles, controlling factors and consequences for water quality. Eur J Soil Sci 58(3):523–546. https://doi.org/10.1111/J.1365-2389.2007.00915.X
Jarvis NJ, Stahli M, Bergstrom L, Johnsson H (1994) Simulation of dichlorprop and bentazon leaching in soils of contrasting texture using the MACRO model. J Environ Sci Health. Part A: Environ Sci Eng Toxicol 29(6):1255–1277. https://doi.org/10.1080/10934529409376106
Jarvis N, Koestel J, Larsbo M (2016) Understanding preferential flow in the vadose zone: recent advances and future prospects. Vadose Zone J 15(12):vzj2016.09.0075. https://doi.org/10.2136/VZJ2016.09.0075
Jégou D, Schrader S, Diestel H, Cluzeau D (2001) Morphological, physical and biochemical characteristics of burrow walls formed by earthworms. Appl Soil Ecol 17(2):165–174. https://doi.org/10.1016/S0929-1393(00)00136-0
Josa R, Gorchs G, Ginovart M, Solé-Benet A (2013) Influence of tillage on soil macropore size, shape of top layer and crop development in a sub-humid environment. Biologia (Poland) 68(6):1099–1103. https://doi.org/10.2478/s11756-013-0250-y
Lawes JB, Gilbert JH, Warington R (1881) On the amount and composition of the rain and drainage-waters collected at Rothamsted. J Roy Agric Soc England
Leue M, Gerke HH (2016) Roughness of biopores and cracks in Bt-horizons assessed by confocal laser scanning microscopy. J Plant Nutr Soil Sci 179(4):529–536. https://doi.org/10.1002/JPLN.201600016
Leue M, Uteau-Puschmann D, Peth S, Nellesen J, Kodešová R, Gerke HH (2019) Separation of soil macropore types in three-dimensional x-ray computed tomography images based on pore geometry characteristics. Vadose Zone J 18(1):1–13. https://doi.org/10.2136/vzj2018.09.0170

Lighthill MJ, Whitham GB (1955) On kinematic waves I. Flood movement in long rivers. In: Proceedings of the royal society of London. Series A. Mathematical and physical sciences, vol 229, no 1178, pp 281–316. https://doi.org/10.1098/RSPA.1955.0088

Luo L, Lin H, Halleck P (2008) Quantifying soil structure and preferential flow in intact soil using X-ray computed tomography. Soil Sci Soc Am J 72(4):1058–1069. https://doi.org/10.2136/SSSAJ2007.0179

Luxmoore RJ, Jardine PM, Wilson PM, Jones GV, Zelazny JR (1990) Physical and chemical controls of preferred path flow through a forested hillslope. Geoderma 46

Luxmoore RJ (1981) Micro-, meso-, and macroporosity of soil. Soil Sci Soc Am J (United States) 45:3(3):671–672. https://doi.org/10.2136/SSSAJ1981.03615995004500030051X

Masciopinto C (2012) Comments on theory for source-responsive and free-surface film modeling of unsaturated flow. Vadose Zone J 11(4):vzj2012.0015. https://doi.org/10.2136/vzj2012.0015

Miller EE, Miller RD (1955) Theory of capillary flow: I. Practical implications. Soil Sci Soc Am J 19(3):267–271. https://doi.org/10.2136/SSSAJ1955.03615995001900030005X

Mohanty BP, Bowman RS, Hendrickx JMH, Van Genuchten MT (1997) New piecewise-continuous hydraulic functions for modeling preferential flow in an intermittent-flood-irrigated field. Wiley Online Libr 33(9):2049–2063. https://doi.org/10.1029/97WR01701

Mooney SJ, Morris C (2008) A morphological approach to understanding preferential flow using image analysis with dye tracers and X-ray computed tomography. Catena 73(2):204–211. https://doi.org/10.1016/j.catena.2007.09.003

Nimmo JR (2010) Theory for source-responsive and free-surface film modeling of unsaturated flow. Vadose Zone J 9(2):295–306. https://doi.org/10.2136/vzj2009.0085

Nimmo JR (2016) Quantitative framework for preferential flow initiation and partitioning. Vadose Zone J 15(2):1–12. https://doi.org/10.2136/vzj2015.05.0079

Nimmo JR, Mitchell L (2013) Predicting vertically nonsequential wetting patterns with a source-responsive model. Vadose Zone J 12(4):vzj2013.03.0054. https://doi.org/10.2136/vzj2013.03.0054

Nimmo JR (2007) Simple predictions of maximum transport rate in unsaturated soil and rock. Water Resour Res 43(5). https://doi.org/10.1029/2006WR005372

Nimmo JR (2012) Preferential flow occurs in unsaturated conditions. In: Hydrological processes, vol 26, Issue 5, pp 786–789. https://doi.org/10.1002/hyp.8380

Nimmo JR (2020) The processes of preferential flow in the unsaturated zone.https://doi.org/10.1002/saj2.20143

Orozco-López E, Muñoz-Carpena R, Gao B, Fox GA (2018) Riparian vadose zone preferential flow: review of concepts, limitations, and perspectives. Vadose Zone J 17(1):1–20. https://doi.org/10.2136/VZJ2018.02.0031

Pagliai M, Lamarca M, Lucamante G (1983) Micromorphometric and micromorphological investigations of a clay loam soil in viticulture under zero and conventional tillage. J Soil Sci 34(2):391–403. https://doi.org/10.1111/J.1365-2389.1983.TB01044.X

Pagliai M, Marca M La, Lucamante G, Genovese L (1984) Effects of zero and conventional tillage on the length and irregularity of elongated pores in a clay loam soil under viticulture. In: Soil and tillage research, vol 4

Perret J, Prasher S, Kantzas A, Langford CH (1999) Three-dimensional quantification of macropore networks in undisturbed soil cores_enhanced reader. Soil Sci Soc Am J

Peters RR, Klavetter EA (1988) A continuum model for water movement in an unsaturated fractured rock mass. Water Resour Res 24(3):416–430. https://doi.org/10.1029/WR024I003P00416

Radulovich R, Solorzano E, Sollins P (1989) Soil macropore size distribution from water breakthrough curves

Rahe TM, Hagedorn C, McCoy EL, Kling GF (1978) Transport of antibiotic-resistant escherichia coli through western Oregon hillslope soils under conditions of saturated flow. American Society of Agronomy, Crop Science Society of America, and Soil Science Society of America, vol 7, no 4, pp 487–494

Reynolds ERC (1966) The percolation of rainwater through soil demonstrated by fluorescent dyes. J Soil Sci 17(1):127–132

Richards LA (1931) Capillary conduction of liquids through porous mediums. J Appl Phys 1(5):318–333. https://doi.org/10.1063/1.1745010

Robert, Blevins L (n.d.) Micromorphology of soil fabric at tree root-soil interface dissertation. Presented in Partial Fulfillment of the Requirements for the Degree Doctor of Philosophy in the Graduate School of the Ohio State University

Ross PJ, Smettem KRJ (2000) A simple treatment of physical nonequilibrium water flow in soils. Soil Sci Soc Am J 64(6):1926–1930. https://doi.org/10.2136/SSSAJ2000.6461926X

Sammartino S, Lissy A-S, Bogner C, Van Den Bogaert R, Capowiez Y, Ruy S, Cornu S (2015) Identifying the functional macropore network related to preferential flow in structured soils. Vadose Zone J 14(10):1–16.https://doi.org/10.2136/VZJ2015.05.0070

Šimůnek J, Jarvis NJ, Van Genuchten MT, Gärdenäs A (2003) Review and comparison of models for describing non-equilibrium and preferential flow and transport in the vadose zone. J Hydrol 272(1–4):14–35. https://doi.org/10.1016/S0022-1694(02)00252-4

Šimůnek J, Wendroth O, Wypler N, Van Genuchten (2001) Non-equilibrium water flow characterized by means of upward infiltration experiments. Wiley Online Libr 52(1):13–24.https://doi.org/10.1046/j.1365-2389.2001.00361.x

Slichter CS (1899) United States geological survey. 19th Annuaul Report, p 295

Su GW, Geller JT, Pruess K, Wen F (1999) Experimental studies of water seepage and intermittent flow in unsaturated, rough-walled fractures. Water Resour Res 35(4):1019–1037. https://doi.org/10.1029/1998WR900127

Tao Y, Zou Z, Guo L, He Y, Lin L, Lin H, Chen J (2020) Linking soil macropores, subsurface flow and its hydrodynamic characteristics to the development of Benggang erosion. J Hydrol 586:124829. https://doi.org/10.1016/J.JHYDROL.2020.124829

Tokunaga TK, Wan J (1997) Water film flow along fracture surfaces of porous rock. Water Resour Res 33(6):1287–1295. https://doi.org/10.1029/97WR00473

Urbina CAF, van Dam JC, Hendriks RFA, van den Berg F, Gooren HPA, Ritsema CJ (2019) Water flow in soils with heterogeneous macropore geometries. Vadose Zone J 18(1):1–17. https://doi.org/10.2136/VZJ2019.02.0015

van Genuchten MT (1980) A closed-form equation for predicting the hydraulic conductivity of unsaturated soils. Soil Sci Soc Am J 44(5):892–898.https://doi.org/10.2136/SSSAJ1980.03615995004400050002X

Van den Bygaart AJ, Fox CA, Fallow DJ, Protz R (2000) Estimating earthworm-influenced soil structure by morphometric image analysis. Soil Sci Soc Am J 64(3):982–988. https://doi.org/10.2136/SSSAJ2000.643982X

Warner GS, Nieber JL, Moore ID, Geise RA (1989) Characterizing macropores in soil by computed tomography. Soil Sci Soc Am J 53(3):653–660. https://doi.org/10.2136/SSSAJ1989.03615995005300030001X

Watson KW, Luxmoore RJ (1986) Estimating macroporosity in a forest watershed by use of a tension infiltrometer. Soil Sci Soc Am J 50(3):578–582. https://doi.org/10.2136/SSSAJ1986.03615995005000030007X

Groundwater Vulnerability Assessment Using SINTACS-AHP Model and GIS in Raipur City

Indrajeet Sahu, A. D. Prasad, and Ishtiyaq Ahmad

1 Introduction

A healthy civilization is regarded as having water as its basic component. Additionally, it is in charge of supporting sustainable economic growth. In many regions of the world, groundwater is an extremely valuable supply of drinking water (Ramesh and Elango 2012). The majority of places have significant rates of population increase, unplanned urban area growth, industry, and inadequate sewage infrastructure. The groundwater is prone to contamination, and once it has been contaminated, it is extremely challenging to clean it up. Urban groundwater pollution is mainly because of effluent coming out from the industries and leakage of sewage and septic tank present in the city. Whereas in outskirt of the city where vegetation is there the factor responsible for pollution is over and regular use of fertilizer and pesticides in framing, all these are resulting in groundwater pollution. Therefore, preventing contamination should be the fundamental of sustainable groundwater management. How easily or difficultly pollution or contamination can travel from the land surface to the producing aquifer is determined by the groundwater's vulnerability (Kazakis and Voudouris 2015). Groundwater Vulnerability Assessment (GVA) is required for proper water availability and to reduce negative effects. As a result, one of the most significant scientific areas today is the assessment of groundwater sensitivity to contamination. To assess aquifer vulnerability, numerous methods have been developed, process-based, statistical, and overlay/index approaches are some of these methods. The majority of overlay and index-based groundwater vulnerability mapping approaches include ratings and attributes of major elements (such as contaminant characteristics,

I. Sahu (✉)
Department of Civil Engineering, Vardhaman College of Engineering, Hyderabad, India
e-mail: iamindraa14@gmail.com

A. D. Prasad · I. Ahmad
Department of Civil Engineering, National Institute of Technology Raipur, Raipur, India

K. K. Singh and C. S. Prasad Ojha (eds.), *Sustainable Management of Land, Water and Pollution of Built-up Area*, Society of Earth Scientists Series,
https://doi.org/10.1007/978-3-031-56176-4_10

recharge rates, soil/aquifer properties, depth to the water table, land use, and management techniques) that affect the transport of pollutants from the ground surface to an aquifer.

Vulnerability assessment, which depends on the aquifer's geological and hydrological characteristics, aims to assist in the development of useful guidelines and policies to manage sustainable resources. The most common method for preserving groundwater quality also involves figuring out how susceptible it is to contamination. The capacity of vulnerability assessments to identify regions that are more easily contaminated than others as a result of anthropogenic activity on or near the earth's surface has received widespread recognition (Babiker et al. 2005). The overlay/index method has been the most commonly used of these major techniques for GVA and aquifer sensitivity on a large scale. The GVA overlay/index approach for this investigation, SINTACS method was utilized. Recent research has employed the SINTACS approach and the GIS environment to estimate the vulnerability of groundwater and found to be more precise than DRASTIC method (Cusimano et al. 2004; Al-Kuisi et al. 2006; Leal and Silva 2012; Kumar et al. 2013; Guastaldi et al. 2014; Murthy et al. 2015; Busico et al. 2017; Noori et al. 2019; Malik and Shukla 2019). With spatial data and hydrogeological data, the GVA can be performed it is more effective yet economical. The model's benefit is that all seven parameters are combined into a single layer, and GIS makes it possible to use a variety of data processing, spatial analysis, and georeferencing techniques to create a precise vulnerability map (Al-Kuisi et al. 2006; Leal and Silva 2012; Kumar et al. 2013). The model will be more effective if the weight and rating which the model used to depict the result will be optimized as per the local hydrogeological scenario (Noori et al. 2019). This study's objective is to assess the groundwater sensitive areas of the city of Raipur using the Standard SINTACS Vulnerability Index (SVI) and Modified SINTACS Vulnerability Index (MSVI) techniques in a geographic information system (GIS) system. The Multi-Criteria Decision Making (MCDM) tool Analytical Hierarchy Process (AHP) was used to modify SINTACS by redefining the weightage in accordance with the hydrological system of the study area (Santhosh and Sivakumar Babu 2018; Soyaslan 2020; http://cgwb.gov.in/index.html). Sensitivity Analysis (SA) was carried out in order to classify the impact of different parameters on GVA and to insure the consistency of the final judgment.

2 Study Area

In Chhattisgarh State, India, Raipur serves as both the district and divisional capital. The study area (Fig. 1) encompasses a region of about 226 km^2 and is located between the longitudes of 81° 35′ and 81° 40′ and the latitudes of 21° 10′ and 21° 20′. The climate in Raipur is tropical, dry, and humid, with mild temperatures that last during the year. The low temperature is 10 °C, and the high temperature can reach 48 °C. According to IMD-Indian Metrological Department, Raipur, the area receives 1460 mm of rain on average, and the water table is reported to be 5–30 mbgl deep

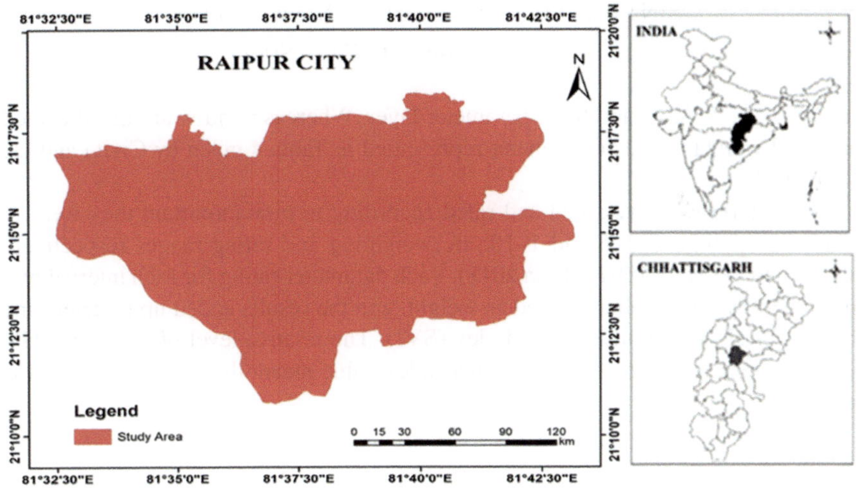

Fig. 1 Study area

(CGWB-Central Ground Water Board) (http://cgwb.gov.in/index.html). The average water level trend for pre and post monsoon periods (during the last ten years) shows no notable change. The city's primary surface water supply comes from the Kharun River, which runs from the city's southern west to eastern side. The research region has established soil types and is broadly categorized into two significant geomorphic units. Those are the Seonath-Mahanadi Alluvial Plain, a Proterozoic shale-limestone-dolomite region. Due to the presence of limestone the city has many cements plant operating here. Population of the city is 20 lakh as per 2011 census.

3 Materials and Methodology

3.1 SINTACS

Early in the 1990s, the SINTACS method was created in Italy to expand the mapping to a broader scale due to the country's extensive hydrogeological diversity. (Aller et al. 1987; Civita and De Maio 1997, 2007). The SINTACS method generally used seven hydrogeological parameters to assess groundwater vulnerability (GV). The parameters were considered as (S)-depth to water table, (I)-net recharge, the (N)-impact of vadose zone, (T)-soil media, (A)-aquifer media, (S)-topography, and (C)-hydraulic conductivity (Eq. 1). The input information and data, such as borehole data, meteorological, hydrological, geology, soil data, topography map, contour map, and lithology data, were used to develop the GIS database (Sener and Davraz 2013).

$$\text{SINTACS(Vi)} = \text{SrSw} + \text{IrIw} + \text{NrNw} + \text{TrTw} + \text{ArAw} + \text{CrCw} + \text{SrSw} \quad (1)$$

here, S, I, N, T, A, C, and S are parameter layer. Whereas r and w are the allocated rates and weights to each parameter represented in Table 4 given by Civita and De Maio (1997).

The criteria were weighted and rated according to how important they were in relation to pollution. 1–5 and 1–10, are weighting and rating ranges respectively (Civita and De Maio 1997; Maria 2018). Each parameter rating for each interval was multiplied by a multiplier defined as weight, and the results added up to determine the final SINTACS Vulnerability Index (SVI). The relative level of a region's GV was represented by this index. A higher index value meant that contamination was more likely.

3.2 *Modified SINTACS*

Groundwater susceptibility for a particular contaminant or hydrogeological characteristic cannot be determined using the standard SINTACS model. Sector-specific research may benefit from parameter selection and addition, although the best weights and rates for the selected parameters are frequently ignored. When the selected parameters are connected, as in the case of groundwater depth and the effect of the vadose zone or slope and net recharge, this becomes even more important. Therefore, eliminating redundancy in factor weights and rates is essential for effective model predictions. This limitation can be overcome by using MCDM methods.

The modified SINTACS method's weights for each parameter were determined using the AHP approach, and the findings were used in combination with GIS techniques to assess GV. In order to analyse the GV, the modified SINTACS is utilized with the AHP, a MCDM approach to change the weights of the input parameters. The weights of the particular characteristics were established based on their relevance to the assessment of GV (Noori et al. 2019; Saaty 2003). After the hierarchy has been created, all seven parameters were subjected to numerous pair-wise comparisons using the nine-level standardized comparison scale Table 1. This makes the decision-makers to assess the objective separately, and also simplify the process in decision making (Maria 2018). The comparison is enumerated using a Saaty's scale where the no. signifies the importance, 9 (extremely more important) to 1(Equal importance) and vice versa (Saaty 2003) Table 1 (Table 2).

In choosing the comparison criterion, the comparison matrix's principal Eigenvalue and associated normalized Eigenvector are crucial (Dengiz et al. 2015) Table 3. The consistency of the matrix of order "n" was then evaluated; if it fell short of an established threshold, the evaluations were reexamined (Sulistio et al. 2018).

The consistency index (CI), was calculated using Eq. 2 as given below:

Table 1 Pair-wise comparison for all seven parameters

Parameters		SD	N	I	C	A	T	S
Water table depth (SD)	SD	1	1.5	2	3	5	5	7
Unsaturated (Vadose zone) (N)	N	0.667	1	3	2.5	5	5	7
Net recharge (I)	I	0.5	0.333	1	2	3	3	5
Hydraulic conductivity (C)	C	0.333	0.4	0.5	1	2	2.5	4
Aquifer media (A)	A	0.2	0.2	0.333	0.5	1	2	3
Soil media (T)	T	0.2	0.2	0.333	0.4	0.5	1	2
Topographic slope (S)	S	0.143	0.143	0.2	0.25	0.333	0.5	1

Note Saaty's scale of pairwise comparison matrix is used to compare (Saaty 2003)

Table 2 Normalized pairwise matrix, each column is divided by their respective sum (Table 1), and criteria weight is obtained by averaging the row

	SD	N	I	C	A	T	S	Criteria weight
SD	0.329	0.397	0.271	0.311	0.400	0.263	0.241	**0.316**
N	0.219	0.265	0.407	0.259	0.400	0.263	0.241	**0.294**
I	0.164	0.088	0.136	0.207	0.240	0.158	0.172	**0.167**
C	0.110	0.106	0.068	0.104	0.160	0.132	0.138	**0.117**
A	0.066	0.053	0.045	0.052	0.080	0.105	0.103	**0.072**
T	0.066	0.053	0.045	0.041	0.040	0.053	0.069	**0.052**
S	0.047	0.038	0.027	0.026	0.027	0.026	0.034	**0.032**

Table 3 Consistency ratio analysis, each column (Table 1) is multiplied by their respective criteria weight and weighted sum are obtained by adding the row

	SD	N	I	C	A	T	S	Weighted sum value	Criteria weight	Ratio w/c
SD	0.316	0.440	0.333	0.350	0.360	0.262	0.225	2.287	**0.316**	7.235
N	0.211	0.294	0.500	0.292	0.360	0.262	0.225	2.143	**0.294**	7.302
I	0.158	0.098	0.167	0.233	0.216	0.157	0.161	1.19	**0.167**	7.146
C	0.105	0.117	0.083	0.117	0.144	0.131	0.129	0.827	**0.117**	7.087
A	0.063	0.059	0.056	0.058	0.072	0.105	0.097	0.509	**0.072**	7.066
T	0.063	0.059	0.056	0.047	0.036	0.052	0.064	0.377	**0.052**	7.19
S	0.045	0.042	0.033	0.029	0.024	0.026	0.032	0.232	**0.032**	7.208
								Eigenvector λmax =		7.302

$$\mathrm{CI} = (\lambda\mathrm{max} - \mathrm{n})/\mathrm{n} - 1$$
$$\mathrm{CI} = (7.302 - 7)/6 = 0.503 \quad (2)$$

where, n is matrix of order 'n', λmax is the largest Eigenvalue of the matrix.

Table 4 SINTACS parameters weight and ratings for GVA

Parameters	Sub-parameter	Rating	Weight	Index rating	AH weight	AHP Rating
Depth to water table (m) (Sd)	0–10	10		50	**0.316**	3.16
	Oct–20	7	5	35		2.22
	20–30	5		25		1.58
	30–40	3		15		0.94
	> 40	1		5		0.31
Net recharge (I)	(VH) Very high	10		40	**0.294**	2.94
	(H) High	9	4	36		2.646
	(M) Medium	8		32		2.352
	(L) Low	6		24		1.764
	(VL) Very low	2		8		0.588
Impact of vadose (N)	Stromatolite dolomite- limestone with sandstone	9		45	**0.167**	1.503
	Stromatolite dolomite-limestone	8	5	40		1.336
	Sandy clay loam	3		15		0.501
	Laterite	2		10		0.334
	Clay loam	1		5		0.167
Soil media (T)	Sandy loam	4		16	**0.117**	0.468
	Gravelly sandy Clay loam	3	4	12		0.351
	Sandy clay loam	3		12		0.351
	Clay sandy loam	2		8		0.234
	Clay loam	2		8		0.234
	Clay	1		4		0.117
Aquifer media (A)	Stromatolite dolomite- limestone with sandstone	10		30	**0.072**	0.72
	Stromatolite dolomite-limestone	9	3	27		0.648
	Laterite	1		3		0.072
Hydraulic conductivity (C) m/day	0.8	9		27	**0.052**	0.468
	0.6	8	3	24		0.416
	0.000864	1		3		0.052
Slope (S)	Very low (VL)	10		20	**0.032**	0.32
	Low (L)	9	2	18		0.288
	Moderate (M)	8		16		0.256

(continued)

Table 4 (continued)

Parameters	Sub-parameter	Rating	Weight	Index rating	AH weight	AHP Rating
	High (H)	7		14		0.224
	Very High (VH)	6		12		0.19

Source for Rating and weight Civita and De Maio (1997)

CR is calculated as CI/RI = 0.038, where RI is the random index. Saaty (1980) gave the average of the random index (value for matrices of order 1–10) (Saaty 2003). RI for seven orders of the matrix is 1.32. The value of CR must be less than 0.1 in order for the matrix to be consistent; if it does not meet this requirement, the comparison's results were reexamined. The data is reliable because the study's CR (0.038) is less than 0.1, and it will be evaluated to determine the modified weight of the parameters.

3.2.1 Sensitivity Analysis

An essential component of the study that aids in estimating more accurate outcomes is sensitivity analysis (SA). To categories the impact of different parameters on GVA, AHP-SA has been used. To guarantee the consistency of final decisions, such decision-making procedures must be subjected to sensitivity analysis. Sensitivity analysis allows for the visualization of possibilities, which is useful for observing how modifying the criteria would affect the final alternative rank. It also aids in determining how many changes are brought about by slight variations in the weights of the criteria (Chen et al. 2013). Assessing a parameter's significance in relation to other factors and GV, applied weights and rating for calculating Vulnerability Index (VI) will differ as a modification of the standard SINTACS for the specific study area (Al-Shatnawi et al. 2016).

Input parameters separately rated by SVI rates and AHP weights. To evaluate Modified SINTACS-AHP Vulnerability Index (MSVI) modified weights and rating were used to obtain the result using the Eq. 3:

$$\text{Modified SINTACS - AHP} = \text{SySa} + \text{IyIa} + \text{NyNa} + \text{TyTa} + \text{AyAa} + \text{CyCa} + \text{SySa} \tag{3}$$

where y and a are the rates and AHP weights allotted to parameters. Modified weights and ratings obtained from the AHP analysis are shown in Table 4 (Fig. 2).

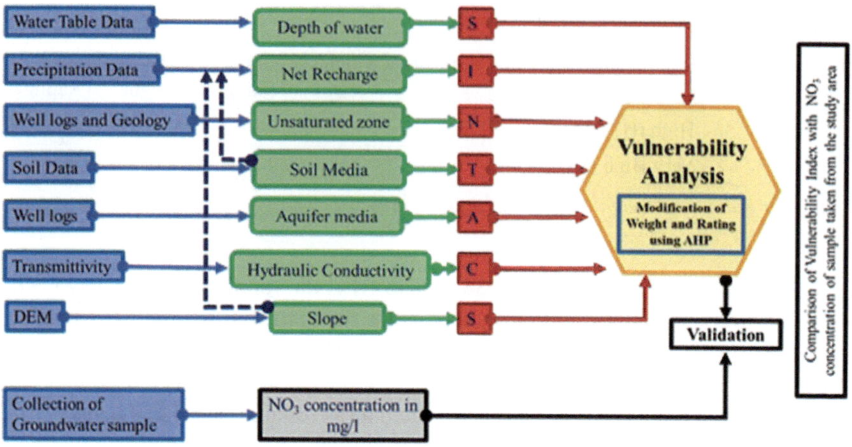

Fig. 2 Methodology

3.3 *SINTACS Parameter*

3.3.1 Water Table Depth (S-Parameter)

The water table depth is defined as the distance from the ground's surface from groundwater level. The depth to the ground water table surface was interpolated from the ground level point data using the Inverse Distance Weighted (IDW) method. The distance between the surface and the saturation zone, which affects the contaminant's migration, is known as water table depth, (Fig. 3). The groundwater level layer is created in the current study using data on groundwater for 2017 obtained from the Central Ground Water Board (CGWB) (Fig. 3). The water is between three and forty-two meters deep (mbgl) Table 4. Then these values are reclassified into ranges and ratings using the "Reclassify" option in the "Spatial Analyst" tool in GIS software according to the rating given by the SINTACS method. The groundwater depth of city comes under 3–40 m with average of 16 m in most of the region.

3.3.2 Net Recharge (I-Parameter)

Net recharge is the quantity of surface water as a result of precipitation that infiltrate from the ground's surface and reaches the saturation zone (Eq. 4). To create this map, various details including slope, precipitation, and soil permeability are required. For the research area, average annual rainfall data from 2010 to 2018 were received from the India Meteorological Department (IMD) in Raipur. Data on soil permeability from the CGWB for the years 2008 and 2016 were also collected. A net recharge layer with a depth of 0 to 250 mm was created using these data and the GIS environment in combination with overlay analysis (Sener and Davraz 2013). The maps used in

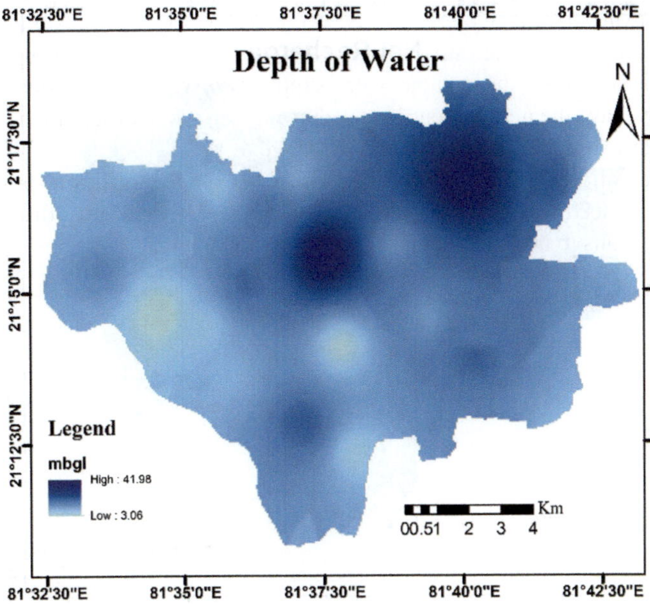

Fig. 3 Water table depth

this overlay analysis are in raster format with dimensions as the slope (%), rainfall (mm) and soil permeability (cm/s). To add these three quantities, each of which has different dimensions to get the recharge value, these layers are reclassified (Fig. 4).

$$\text{Recharge value} = \text{Soil permeability} + \text{Rainfall} + \text{Slope\%} \quad (4)$$

3.3.3 Impact of Vadose Zone (N-Parameter)

The aquifer is replenished/recharged by water that initially percolates from the vadose zone (Unsaturated Zone). Aquifer is space where the groundwater is found or in other term it gets stored here. As a result, both the quality and quantity of groundwater are significantly influenced by the vadose (unsaturated) zone. Using borehole data acquired from GSI and CGWB, the current study developed the effect of the unsaturated zone (Fig. 5), it is given a weight of 5. Mainly five type of unsaturated zone type are found in the city with major portion of sandstone and limestone.

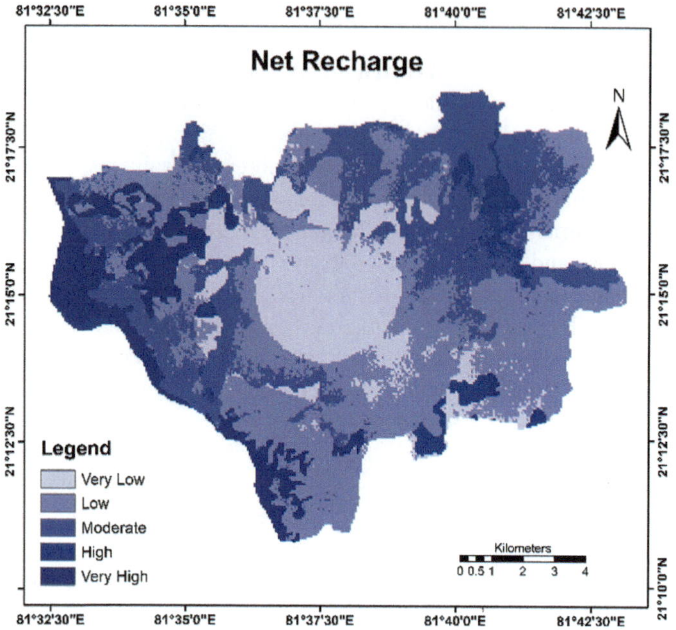

Fig. 4 Net recharge

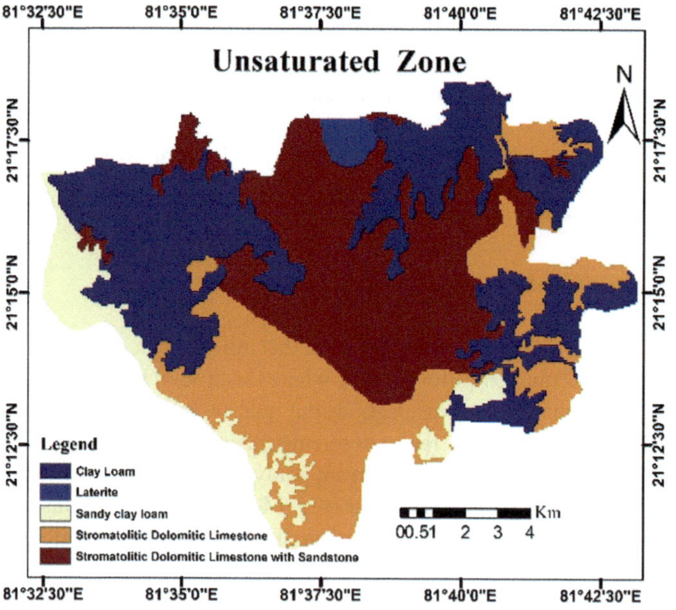

Fig. 5 Impact of vadose zone

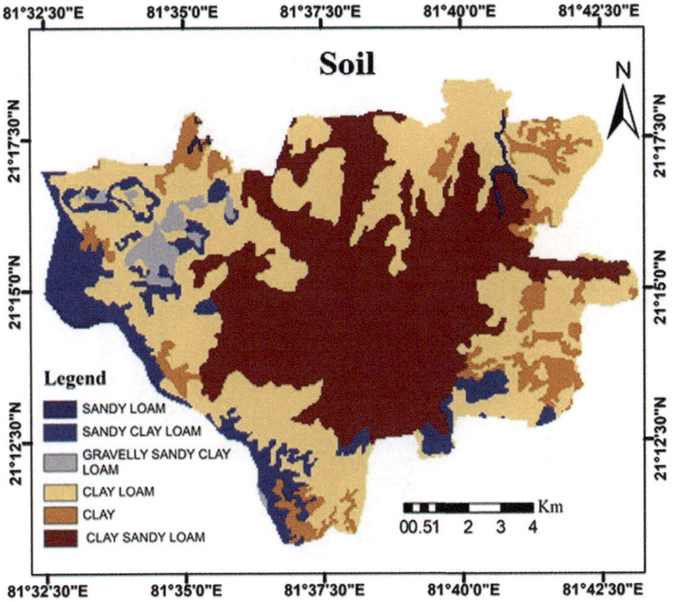

Fig. 6 Soil

3.3.4 Soil Media (T-Parameter)

One of the important factors in assessing groundwater contamination is the soil (Leal and Silva 2012). It descends from the upper stratum to a depth of around 6 feet. The frequency of infiltration is determined by the soil's characteristics. The texture of the soil affects permeability. For the vertical downward migration of contaminated goods, soil serves as a portal to groundwater and affects the vulnerability. In the current study, soil maps were created using data acquired from the Chhattisgarh Infotech Promotion Society (CHIPS), Raipur and from the National Bureau of Soil Survey and Land Use Planning (NBSS-LUP) (Fig. 6). There are several types of soil in the area, including clay, clay loam, and clay sandy loam. The soil parameter is assigned a weight of four (Table 4). Major area of the city is having gravelly sandy clay mainly in the central part of the study area.

3.3.5 Aquifer Media (A-Parameter)

An aquifer is a layer below the surface that comprises water-bearing sediments, permeable, unconsolidated rocks, or rock fragments from which groundwater can be drawn using a water well. An aquifer has the capacity to deliver adequate water to the spring or well. Anthropogenic activities have an impact on aquifers, and aquifer media composition largely determines the extent of geogenic contamination (Corniello et al. 2004; Umar et al. 2009; Srinivasamoorthy et al. 2011). Limestone and dolomite make

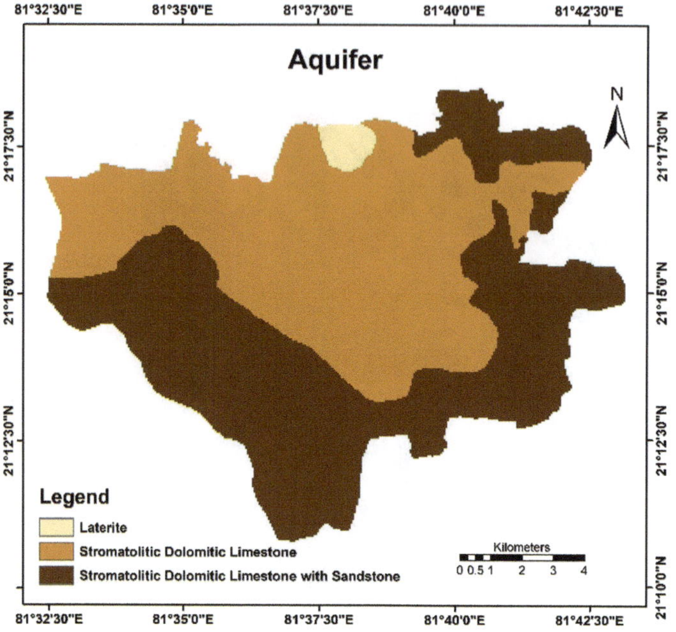

Fig. 7 Aquifer

up the majority of the region's aquifer system. The area's good aquifer is formed by the limestone of the Chandi formation, which is cavernous. Additionally, the alluvium blankets that lines the main rivers serves as a good groundwater storage area. As demonstrated by the CGWB Raipur data used to generate the aquifer media layer (Fig. 7). According to SINTACS standards, the rating of aquifer media (A-Parameter), varies between value 1 for Laterite and value 10 for Stromatolitic dolomitic limestone with sandstone. the weight given to the aquifer Media parameter is 3. Aquifer is of three type presents in the study area with limestone and sandstone covering the major portion.

3.3.6 Hydraulic Conductivity (C-Parameter)

Hydraulic conductivity is a measure for fluid flow rate. Additionally, it controls the level of pollution by regulating hydraulic conductivity, fluid velocity, and increases in the rate at which contaminants are transported, leading to an increase in aquifer sensitivity (Shirazi et al. 2013). The CGWB well's hydraulic conductivity ranges from 0.02 to 1.1 m/day in the Chandi Formation and from 0.5 to 40 lps in the Chhattisgarh Formation. The CGWB data was used to build the hydraulic conductivity layer for the research area (Fig. 8). According to Civita and De Maio (1997), the values are categorized into three zones, with ratings as 1, 8, and 9 and weight assigned for SVI is 3. Most of Southern central part is having conductivity of 0.80 m/day.

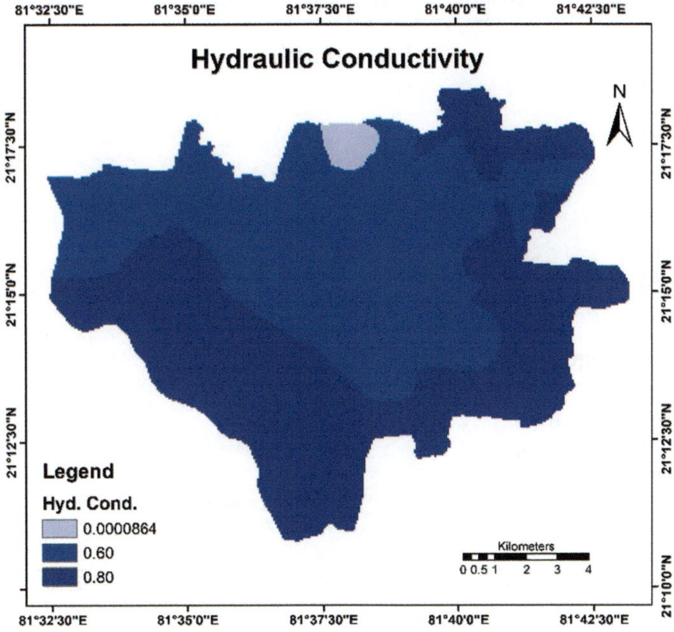

Fig. 8 Hydraulic conductivity

3.3.7 Topography (S-Parameter)

The measurement of a region's steepness is its slope. Compared to fields with high slopes, the low slope area aids in water retention, giving water extra time to infiltrate. Low-slope areas are more prone to groundwater pollution and have larger contamination potentials (Neshat and Pradhan 2017). Using the Digital Elevation Model (DEM) of CartoSAT with a resolution of 30 m, the topographical slope is determined. Figure 9, represents the slope-gradient of the area, which is from the center to the outside part, or from the central section to the outer boundary, where the elevation is greater and gradually lower. The slope is in between 199 and 256 m over all variation of about 50 m eastern part is having low slope as river flows from there whereas the central part slope is on higher side than other.

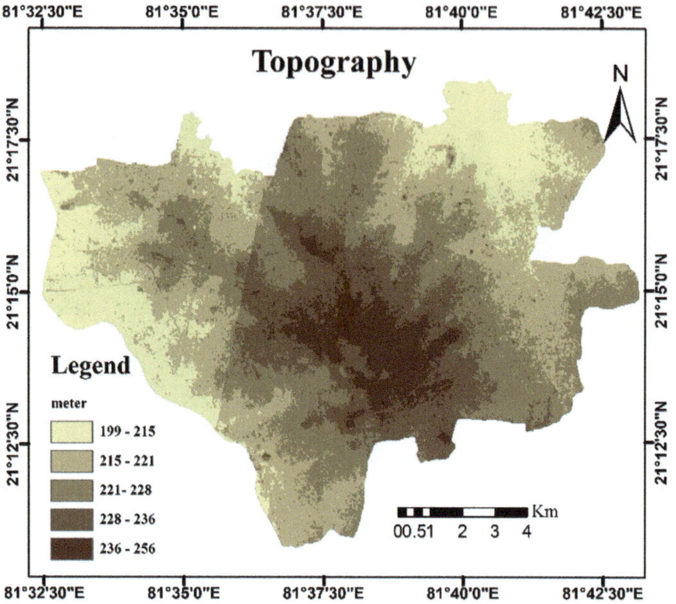

Fig. 9 Topography

4 Results

4.1 SINTACS

For the study region, the SINTACS index numbers ranged from 84 to 215. The acquired vulnerability map shows how sensitive different aquifer contamination concentrations are (Fig. 11). The vulnerability index of the study area evaluated by the methods applied is presented in the form of a map for visualization. The SVI result is standardized and divided into five classes. SVI-based categories are labelled as, Very-low, Low, Moderate, High, and Very-high Vulnerability presented in Table 6 (Busico et al. 2017; Sener and Davraz 2013; Saaty 2003). The effectiveness of the models was analyzed using field investigation (Table 5).

4.2 Modified SINTACS-AHP

AHP an MCDM method was used to modify the SINTACS method as MSVI, based on the local influencing factors, its weight and rating are modified for better prediction of vulnerable zones. The modified AHP weight and rating are shown in Table 4. Using Eq. (3), a similar process was performed to determine the vulnerable zone.

Table 5 Critical and sensitivity degree of coefficient for rank reversal

Difference in weight of the alternative (k)	Criticality degree of criterion (Dk)	Sensitivity coefficient of criterion (Sk) = 1/Dk
Case V1	–	–
(Case V2) 0.045	61.85	0.016
(Case V3) 0.023	**7.69**	**0.13**
(Case V4) 0.05	42.79	0.023
(Case V5) 0.034	37.46	0.026

Table 6 Distribution of study area into different class of the vulnerability index

S. No.	Class	SINTACS vulnerability index		SINTACS-AHP vulnerability index		Percentage change (%)
		Area (%)	Area (km^2)	Area (%)	Area (km^2)	
1	Very low	5.42	12.24	8.56	18.83	57.93
2	Low	16.38	37.02	20.14	44.31	22.95
3	Moderate	36.4	82.26	32.2	70.84	11.54
4	High	24.54	55.46	26.68	58.7	8.72
5	Very high	14.26	32.23	12.42	27.32	12.9

The obtained vulnerability index value ranges between 15.38 and 38.06. Modified SINTACS-AHP vulnerability Index map shows, 8.56, 20.14, 32.20, 26.68, and 12.42% area falls under Very low, low, moderate, high and very highly vulnerable Table 6, MSVI classes are shown in Figs. 11 and 12.

4.2.1 Sensitivity Analysis

In this case, simulation of sensitivity analysis is carried out by making the pairwise comparison between different alternative cases i.e., Case V1 (Depth of water), Case V2 (Net Recharge), Case V3 (unsaturated zone), Case V4 (soil media), and Case V5 (Topography) with each other, and then observing the rank order due to any changes. The difference in weight of the alternative (k) is found out, after that the criticality degree of the criterion (Dk) is determined, which is the minimum percentage by which the current value will change, causing the current ranking of the alternatives to change. The reciprocal of its criticality degree is thus the sensitivity coefficient of the criterion (Sk). The Smallest Dk value or highest Sk value results in a sensitive parameter and can lead to rank reversal (Table 5). The rank reversal occurs to the first and third ranks (Depth of water and Unsaturated zone) when the topography aspect is given high importance. Another rank reversal occurs between first and second when priority is given to the unsaturated zone (Fig. 10). The accumulated result shows that Case V3 (Unsaturated zone) is having a higher rank and will be best suited for the condition prevailing to the study area. Overall, it can be said that the judgement is consistent and reliable based on the sensitivity analysis.

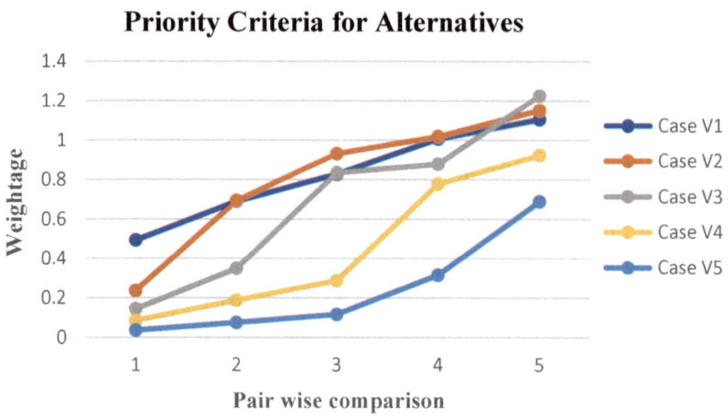

Fig. 10 Priority/rank criteria for alternative cases using sensitivity analysis

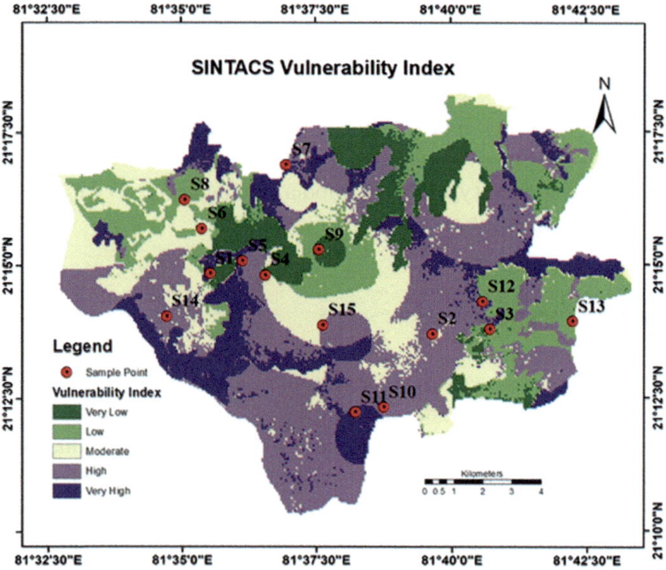

Fig. 11 SVI along with Nitrate sample point Location

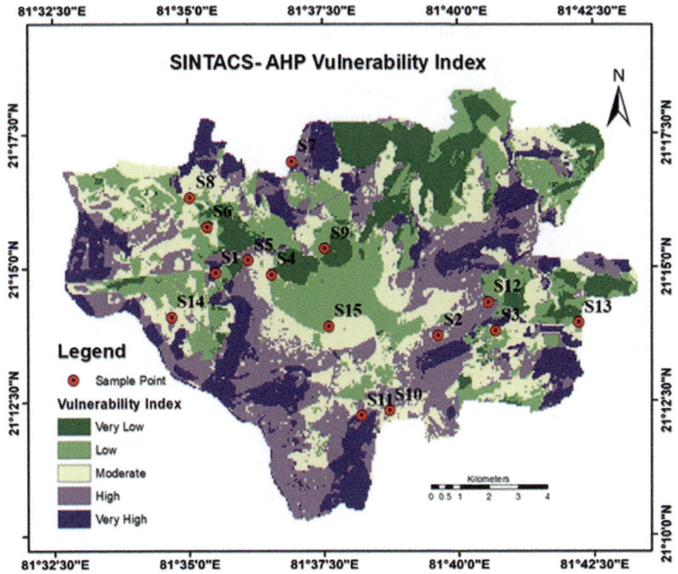

Fig. 12 MSVI along with Nitrate sample point Location

5 Validation and Discussion

The quantity of nitrate conc. measured in the area (one of the main groundwater contaminants) validated the delineated groundwater sensitive zones (Ramesh and Elango 2012; Kazakis and Voudouris 2015; Kumar et al. 2013; Busico et al. 2017). The model's reliability in assessing vulnerability to groundwater contamination was studied by examining the relationship between the groundwater nitrate levels and the Vulnerability index. Validation is done using a simple overlay technique, where the nitrate concentration location layer is overlayed to the vulnerability index layer. Then the value of both layers is observed. If the value range of nitrate concentration is high and the value of the vulnerability class of that point is under the high zone, then a strong relationship between the two is established (Table 7).

In order to determine the nitrate concentration, 15 groundwater samples were collected from the scattered wells situated in the study area's identified groundwater vulnerable zones. These samples were then tested using the 4500-NO_3-B Ultraviolet Spectrophotometric Screening Method (Figs. 11 and 12). The results of the chemical analysis show that NO_3^- values for all groundwater samples range from 8 mg/l to 84 mg/l. Since 45 mg/l is the highest-desired and maximum allowable nitrate limit that is safe for drinking, nitrate concentrations beyond that level in groundwater are dangerous for human consumption (IS:10500) Drinking water standards (https://doi.org/10.2105/SMWW.2882.089). The concentration of nitrate is between 15 and 40 mg/l in the moderately vulnerable zone (m), and between 40 and 125 mg/l in the highly vulnerable zone (h).

Table 7 Comparison of the concentration at various sample locations with the corresponding vulnerability index of sample point location

Location		Sample number	Nitrate concentration (mg/l)	Range-of nitrate conc	SINTACS SINTACS vulnerability Index	SINTACS -AHP vulnerability Index
(Latitude) E	(Longitude)N					
81.588	21.252	S1	48.12	h	H	H
81.605	21.252	S2	65.34	h	H	H
81.598	21.256	S3	14.20	l	M	M
81.580	21.275	S4	10.86	l	L	L
81.622	21.260	S5	32.45	m	H	M
81.642	21.212	S6	20.64	m	M	M
81.633	21.210	S7	52.36	h	H	H
81.673	21.244	S8	28.01	m	L	M
81.574	21.239	S9	08.04	l	L	L
81.623	21.237	S10	24.46	m	H	M
81.633	21.210	S11	68.26	h	H	H
81.673	21.244	S12	84.34	h	H	H
81.701	21.238	S13	12.52	l	L	L
81.574	21.239	S14	40.62	m	M	H
81.623	21.237	S15	10.34	l	M	L

Note 1. Nitrate Concentration (h = high, m = moderate, l = low). 2. Vulnerability Index (H = high, M = moderate, L = low)

As there are no acknowledged geological nitrate sources in the area and the majority of the area with high concentrations of nitrate is comprised of agricultural land, fertilizers are likely the main source of nitrate in groundwater. The effectiveness of the used methods SVI and MSVI are evaluated utilizing nitrate concentration value of the sample point location indicated in Table 8. The evaluation of the results shows that MSVI is more accurate than traditional SVI.

Table 8 Accuracy of the applied methods using Nitrate concentration of sample point location

	SINTACS	Modified SINTACS-AHP
Nitrate sample (which does not have similarities)	S5, S8, S10, and S15	S3, and S14
Accuracy% (similarities/total sample)	**73.4**	**86.7**

6 Conclusion

The SINTACS method which is overlay index method has been modified as MSVI method for assessing GV, in conjunction with GIS techniques. The AHP technique has been used to quantify the weight of all the parameters used in the Modified SINTACS method as per hydrogeological setting. Using SINTACS, AHP, and GIS together offers a highly precise way for figuring out how sensitive groundwater is to contamination. The SVI and MSVI for the research region, respectively, vary from 84 to 215 and 15 to 38. From the study region's subsequent for SVI and MSVI map, it is evident that the majority of the area 36% and 32.20%, respectively lies in the moderate vulnerability zone. The lower southern region, which is populated and has agricultural land, as well as the lower western region, which has important industries, are the main areas of the region that are in the extremely high vulnerability zone, accounting for 14.26% for SVI and 12.42% for MSVI of the entire region. Settlement is what causes this high vulnerability.

In addition, to analyse the model's efficacy in determining vulnerability to groundwater pollution, the link between the SVI, MSVI, and groundwater nitrate levels was examined using the Ultraviolet Spectrophotometric Screening Method. When Nitrate is compared to the VI of the study area at particular location four samples show dissimilarities for SVI and for MSVI the dissimilarities is for only two samples. Validation shows that the accuracy of SVI is around 73.4% whereas the MSVI method it is around 86.7%. The modified method is 13.3% more effective in evaluating GVA of Raipur city than the standard method.

The annual net recharge, which is 250 + mm, is the main cause of the extremely high vulnerability, with a high rate near industrial areas (chemical pollutant and wastewater) and the agricultural land (use of pesticides for farming) are the major reason for high VI of those particular regions. In addition, the Unsaturated/vadose zone is also a factor contributing to vulnerability major area has dolomitic, limestone with sandstone making it simpler for water to infiltrate into the ground. The study contributes to effectively analyzing GV by modifying rating and weight as per the hydrogeological settings of the area. The constraint can be the unavailability of accurate data set and some more parameters which can influence the vulnerability can be introduced as a modification along with this standard seven parameters. The scientific value added to the present study is modifying rating and weight utilizing AHP in accordance with the hydrogeological conditions of the region The current study also gives local authorities a useful tool for managing and controlling their groundwater supplies.

References

Al-Kuisi M, El-Naqa A, Hammouri N (2006) Vulnerability mapping of shallow ground-water aquifer using SINTACS model in the Jordan Valley Area Jordan. Environ Geol 50:651–667

Aller L, Bennet T, Leher JH, Petty RJ, Hackett G (1987) DRASTIC: a standardized system for evaluating groundwater pollution potential using hydro-geological settings EPA 600/2-87-035:622

Al-Shatnawi AM, El-Bashir MS, Bani Khalaf RM, Gazzaz NM (2016) Vulnerability mapping of groundwater aquifer using SINTACS in Wadi Al-Waleh Catchment, Jordan. Arab J Geosci 9(1):1–13

Assessment of nitrate concentration using 4500-NO_3-B Ultraviolet Spectrophotometric Screening Method. https://doi.org/10.2105/SMWW.2882.089

Babiker IS, Mohamed MA, Hiyama T, Kato K (2005) A GIS-based DRASTIC model for assessing aquifer vulnerability in Kakamigahara Heights, Gifu Prefecture, central Japan. Sci Total Environ 345(1–3):127–140

Busico G, Colombani N, Cuoco E et al (2017) Modifying SINTACS method to assess groundwater vulnerability and pollution risk to nitrate. Eur Water

Central Ground Water Board (CGWB), Raipur. http://cgwb.gov.in/index.html

Chen Y, Yu J, Khan S (2013) The spatial framework for weight sensitivity analysis in AHP-based multi-criteria decision making. Environ Model Softw 48:129–140

Civita M, De Maio M (1997) SINTACS: Un sistema parametrico per la valutazione e la cartografia della vulnerabilita' degli acquiferi all'inquinamento. Metodologia and Au-tomatizzazione, Pitagora Editrice, Bologna, 60

Civita MV, De Maio M, Fiorucci A (2007) The Italian way to prevent groundwater contamination risk. Water Pract Technol 2(3)

Corniello A, Ducci D, Monti GM (2004) Aquifer pollution vulnerability in the Sorrento peninsula, southern Italy, evaluated by SINTACS method. Geofís Int 43(4):575–581

Cusimano G, De Maio M, Gatto L, Hauser S, Pisciotta A (2004) Application of SINTACS method to the aquifers of Piana di Palermo, Sicily, Italy. Geofís Int 43(4):661–670

Dengiz O, Özyazici MA, Sağlam M (2015) Multi-criteria assessment and geostatistical approach for determination of rice growing suitability sites in Gokirmak catchment. Paddy Water Environ 13(1):1–10

Guastaldi E, Graziano L, Liali G, Brogna FNA, Barbagli A (2014) Intrinsic vulnerability assessment of Saturnia thermal aquifer by means of three parametric methods: SINTACS, GODS and COP. Environ Earth Sci 72(8):2861–2878

Kazakis N, Voudouris KS (2015) Groundwater vulnerability and pollution risk assessment of porous aquifers to nitrate: modifying the DRASTIC method using quantitative parameters. J Hydrol 525:13–25

Kumar S, Thirumalaivasan D, Radhakrishnan N, Mathew S (2013) Groundwater vulnerability assessment using SINTACS model. Geomatics Natural Hazards Risk 4(4):339–354

Leal JAR, Silva FOT et al (2012) Analysis of aquifer vulnerability and water quality using SINTACS and geographic weighted regression. Environ Earth Sci 66(8):2257–2271

Malik MS, Shukla JP (2019) GIS modeling approach for assessment of groundwater vulnerability in parts of Tawa River catchment area, Hoshangabad, Madhya Pradesh, India. Groundwater Sustain Dev 9:100249

Maria R (2018) Comparative studies of groundwater vulnerability assessment. In: IOP conference series: earth and environmental science, vol 118, no 1. IOP Publishing, p 012018

Murthy KR, Dhanakumar S, Sundararaj P, Mohanraj R, Kumaraswamy K (2015) GIS-based modified SINTACS model for assessing groundwater vulnerability to pollution in Vellore District (Part of Palar River Basin), Tamil Nadu, India. In: Environmental management of river basin ecosystems. Springer, Cham, pp 429–453

Neshat A, Pradhan B (2017) Evaluation of groundwater vulnerability to pollution using DRASTIC framework and GIS. Arab J Geosci 10(22):501

Noori R, Ghahremanzadeh H, Kløve B, Adamowski JF, Baghvand A (2019) Modified-DRASTIC, modified-SINTACS and SI methods for groundwater vulnerability assessment in the southern Tehran aquifer. J Environ Sci Health, Part A 54(1):89–100

Ramesh K, Elango L (2012) Groundwater quality and its suitability for domestic and agricultural use in Tondiar river basin, Tamil Nadu, India. Environ Monit Assess 184(6):3887–3899

Saaty TL (2003) Decision-making with the AHP: why is the principal eigenvector necessary. Eur J Oper Res 145(1):85–91

Santhosh LG, Sivakumar Babu GL (2018) Landfill site selection based on reliability concepts using the DRASTIC method and AHP integrated with GIS–a case study of Bengaluru city, India. In: Georisk: assessment and management of risk for engineered systems and Geohazards, vol 12, no 3, pp 234–252

Sener E, Davraz A (2013) Assessment of groundwater vulnerability based on a modified DRASTIC model, GIS and an analytic hierarchy process (AHP) method: the case of Egirdir Lake basin (Isparta, Turkey). Hydrogeol J 21(3):701–714

Shirazi SM, Imran HM, Akib S, Yusop Z, Harun ZB (2013) Groundwater vulnerability assessment in the Melaka State of Malaysia using DRASTIC and GIS techniques. Environ Earth Sci 70(5):2293–2304

Soyaslan İİ (2020) Assessment of groundwater vulnerability using modified DRASTIC-analytical hierarchy process model in Bucak Basin, Turkey. Arab J Geosci 13(21):1–12

Srinivasamoorthy K, Nanthakumar C, Vasanthavigar M, Vijayaraghavan K, Rajivgandhi R, Chidambaram S, Anandhan P, Manivannan R, Vasudevan S (2011) Groundwater quality assessment from a hard rock terrain, Salem district of Tamilnadu, India. Arab J Geosci 4(1–2):91–102

Sulistio H, Waty M, Setiawan MI, Kurniasih N, Ahmar AS (2018) Sensitivity analysis with AHP method: selection of foundation design on bridge Sei Muara Bulan Anak 2. In: Journal of physics: conference series, vol 1028, no 1. IOP Publishing, p 012069

Umar R, Ahmed I, Alam F (2009) Mapping groundwater vulnerable zones using modified DRASTIC approach of an alluvial aquifer in parts of Central Ganga Plain, Western Uttar Pradesh. J Geol Soc India 73(2):193–201

Water and Disasters—A Case Study of India

Rahul Kumar and Anil Kumar Dikshit

1 Introduction

During the last few decades, there has been a continuous rise in disaster events worldwide. The disasters have various origins, have affected millions of people, and result from natural (e.g. tsunami) or technological events (e.g. oil spills). The changing climate and human intervention in the environment with various activities have been significant causes. The disaster can have varied impacts on the environment, the affected population, and the region's infrastructure (Young et al. 2004; Srinivas and Nakagawa 2008).

Centre for Research on the Epidemiology of Disasters (CRED) of the University of Louvain, Brussels, Belgium, has been maintaining an international database, the Emergency Events Database (EM-DAT), since 1988. It is an open-source database that has records of various disasters occurrences across the world and their impacts. It compiles the information from multiple sources, like United Nations agencies, insurance companies, non-governmental organisations, press agencies, and research institutes. CRED has classified the disasters into natural, technological, and complex origin as broad categories and further into sub-groups based on their type of origin (CRED 2020).

Shen et al. (2019) analysed the spatial and temporal data of natural disasters between 1900 and 2015, highlighting severe impacts on all countries worldwide. Especially the large, populated, developed, and fast-developing nations have seen the most adverse effect of the disasters. India ranks 3rd in the number of occurrences, 2nd in fatalities, 2nd in affected population and 5th in recorded damages for worldwide catastrophes.

R. Kumar (✉) · A. K. Dikshit
Environmental Science and Engineering Department, Indian Institute of Technology Bombay, Mumbai 400076, India
e-mail: kumar_rahul@iitb.ac.in

K. K. Singh and C. S. Prasad Ojha (eds.), *Sustainable Management of Land, Water and Pollution of Built-up Area*, Society of Earth Scientists Series,
https://doi.org/10.1007/978-3-031-56176-4_11

Grayman (2011) summarises the overall trend of the disasters in the past decades as:

- The occurrence of disaster events has increased.
- Fatalities have decreased over the past few years.
- Damage to infrastructure and property have been on the rise.
- 90% of all-natural disasters are water-related.

2 Disasters in India

EM-DAT data from January 2000–December 2020 presents the following classifications and count of disasters that occurred in India over the same period (Table 1):

The total number of recorded disasters is 723. Among them, 50% (360) are Natural, and 50% (362) are Technological. Hydrological, 55% and Meteorological, 33%, account for 88% of natural disasters. Floods (178) and storms (85) (including cyclones) comprise 73% of the natural disaster events and have been the most frequent and most destructive natural disasters occurring over the years (Fig. 1).

The population affected by natural disasters was nearly 1120 million (112 Crores) in the last 21 years. On average, nine flood and cyclone events have occurred every year in the same period, affecting almost 424 million (42.4 Crores) people. Although only five recorded drought events, the population affected was around 688 million (68.8 Crores) (Fig. 2).

Table 1 Classification and Count of Disasters

Disaster group	Disaster subgroup	Disaster type	Count of disaster group
Natural	Biological	Epidemic	360
	Climatological	Drought, wildfire	
	Geophysical	Earthquake, mass movement	
	Hydrological	Flood, landslide	
	Meteorological	Storm, extreme temperature	
Technological	Technological	Industrial Accident, Transport Accident, Miscellaneous Accident	362
Complex	Complex	Complex (Food Shortage)	1
Total			**723**

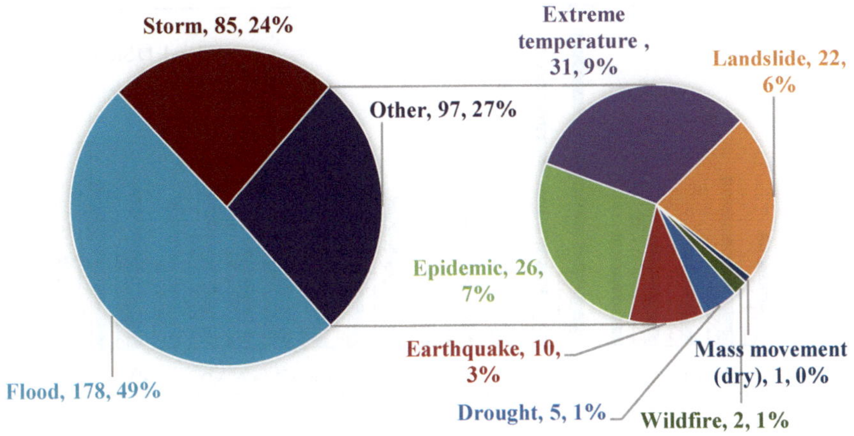

Fig. 1 Distribution of natural disasters by disaster type in India (2000–2020)

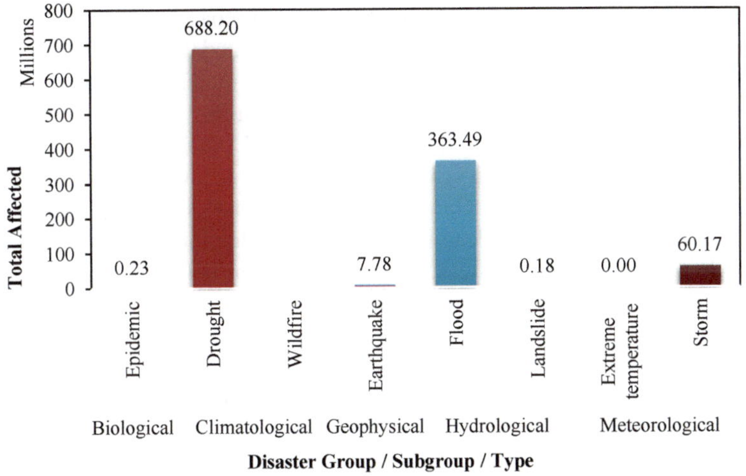

Fig. 2 Population affected by natural disasters in India (2000–2020)

2.1 Water-Related Disasters

Water is one of the primary forces and the engine behind every process that occurs on the earth. Water could be the controlling force or the limiting factor in hydrological, meteorological, biological, etc., phenomenon. Similarly, when it comes to disasters, water is the factor behind all of them, whether directly or indirectly (Grayman 2011; Adikari and Yoshitani 2009).

When analysed, the disaster data of the recent past have proven that India is very prone to water-related disasters due to its geographical location, hot and humid

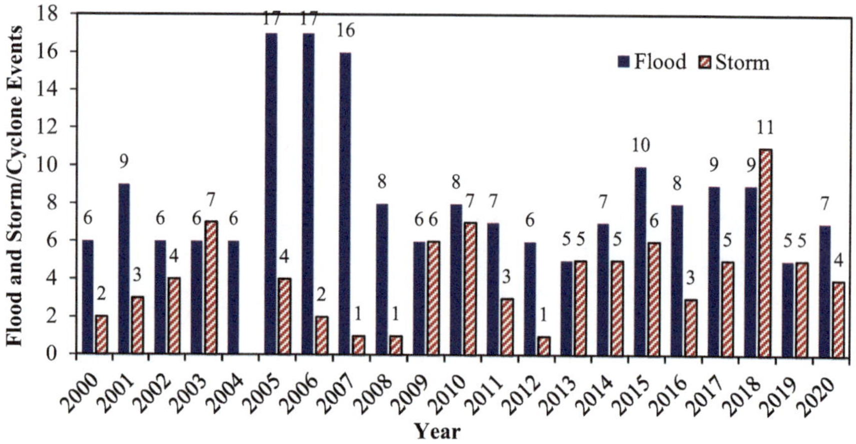

Fig. 3 Flood and storm events in India (2000–2020)

climate. Cyclonic storms are common in the eastern coastal region, where severe cyclones like Fani (2019), Amphan (2020) have devastated the region in recent years. Cyclones result in excessively high winds and heavy rainfall in a short period, leading to landslides in hilly areas and flooding in low lying areas.

Cloud burst and flash floods frequently occurred in Uttarakhand, where the Kedarnath flood (2013) had racked havoc, destroying the whole town of Kedarnath. Floods occur nearly every year in the Gangetic plains of Bihar during the monsoons due to heavy to excessive rainfall, leading to rivers' overflowing and water submerging the low-lying areas along the rivers. Urban floods occur in low-lying areas of cities, mainly due to inadequate drainage systems.

Tsunami (2004), which battered many countries in the Indian Ocean, had left the eastern coastal states of India in despair. Droughts have been common in India's central and southern parts due to the variable and changing rainfall patterns in many regions (CRED 2020).

The analysis of floods and cyclone events shows an average of nine events per year in the past 21 years, affecting millions of people every year (Fig. 3).

3 Impacts of Disasters

Water-related disasters can have direct impacts (e.g.- displacement of people, loss of life and property, struggle for survival, food and water shortage, and environmental damage) and indirect effects (e.g. health problems, loss of livelihood). The economic cost and the cost of mitigation of these disasters require a significant amount of the country's assets and preparedness from the government and relief organisations.

3.1 Impact on Water Sources

The disaster events will have varied impacts on groundwater and surface water. Investigating these impacts on water quality and quantity is essential to understand how we can deal with these situations (Mishra et al. 2021).

Flood. The flood being the most common disaster in India, can significantly impact the dynamics of the water sources. The quality of the water sources like rivers, lakes, etc., can be adversely affected due to the incursion of floodwater. Saha et al. (2020) studied the impact of a massive flood on the Pampa river in Pathanamthitta and Alappuzha, Kerala, India, in 2018. The study was conducted by sampling the river along the stretch of the river after the flood. The author highlights that the flood caused the water to be more alkaline and saline than the pre-flood scenario. The turbidity values exceeded 10 NTU in some cases. There was an increase in dissolved oxygen (DO) > 9 mg/L at some locations and a decrease in the river's heavy metals concentration, which positively impacted the flood's water quality. A possible reason may be the washing of the sediments and dilution due to excess quantity of water. While the water was usable in some cases, the river water required treatment.

Kamalanandhini et al. (2019) inspected the impact of the flood on the groundwater in Kancheepuram, Tamil Nadu, India. The study was conducted by sampling groundwater from various locations in the district. The author reports that groundwater quality was considered good after the flood in 2015 with pH 8.48, total dissolved solids (TDS) 1897 mg/L, total hardness (TH) 580 mg/L and Chlorides 190 mg/L. But over the years, the rainfall in the region has been low, and the water quality has been deteriorating. In 2019 the author reported pH 7.56, total dissolved solids (TDS) 2264 mg/L, total hardness (TH) 910 mg/L and Chlorides 625 mg/L. While these parameters were within permissible limits of drinking water standards IS 10500:2012 during the post-flood period, they exceeded the limits by a good margin in 2019.

Storms. Tropical cyclonic storms are the second most common natural disaster event in India, most prevalent in the Bay of Bengal region. The eastern coast has braved numerous severe cyclones in the past years. Cyclones result from meteorological phenomena that also result in vast amounts of rainfall in the affected region. Mitra et al. (2011) investigated the impact of severe cyclonic storm Aila on the hydrological parameters of the Sunderban delta region before and after the cyclone. The author reported about a 25% change in the salinity and pH while the DO decreased by 5%.

Barik et al. (2017) examined the impact of severe cyclone Phailin on the Chilika lake in Odisha, India. Chilika lake is a brackish water lagoon and is the home of vast biodiversity. The author communicates that the impact of cyclones reduced the salinity. Although the lake's water quality is good, the effect on biodiversity may be harmful.

Tsunami. The tsunami is a result of an earthquake under the sea. The tsunami of 2004 is very well known for the devastation it caused. Its long-term impact can be seen in groundwater contamination in the coastal regions due to seawater incursion into the sources. Chandrasekharan et al. (2008) studied the tsunami's impact on the groundwater of the Nagapattinam District, Tamil Nadu, India. The author collected samples from dug wells and hand pumps /borewells from various locations and compared them with seawater. The author reported that the groundwater pH was 7.9–8.6 in the region. Electrical conductivity [EC] was in the range of 5000–60,000 μS/cm, while the EC of seawater was around 50,000 μS/cm. The saltwater intrusion had made the groundwater unfit for use in agriculture or domestic purposes.

Violette et al. (2009) studied the impact of the tsunami on groundwater in Pondicherry, India. The author reports the increase in salinity due to the inundation of the "Dune aquifer" system present in southern India. The findings suggest that salinity changed from potable levels to EC of 13,000 μS/cm. Groundwater salination was a massive problem as most people depended on groundwater, and it had become unfit for agricultural or domestic purposes.

Singh (2008) investigated the tsunami's impact on groundwater and aquifers at Neill island in Andaman. The author reports that groundwater was the only source of fresh water in the region. Due to saltwater incursion after the tsunami, the EC levels were between 600 and 5600 μS/cm, suggesting a high salinity level.

Drought. The hydrological drought results from prolonged periods of low rainfall, which results in the exploitation of the existing water sources with inadequate replenishment. Kamalanandhini et al. (2021) examined the impact of drought on the water sources in Chengalpatta district, Tamil Nadu, India. The author found that the region's water quality was good and was not significantly impacted by prevailing drought conditions; however, the water sources were under stress due to excessive usage and low replenishment.

Dhawde et al. (2018) investigated the water quality in Pune and Satara districts of Maharastra, India, to determine the impact of the drought. The author reported that the water quality of the water sources was not very severely affected, but high microbial contamination was present, which is detrimental to human health.

3.2 Impacts of Disasters on Water Supply and Sewerage System

The water supply system and sewerage system is affected due to damage to the infrastructure of the treatment plant, the distribution system, the contamination of the source or various other factors (PAHO 2002), as shown in Fig. 4. The classification and scaling of the multiple impacts suggest that scarcity of usable water can arise due to a disaster when the regular water supply is disrupted. Earthquakes, floods and cyclones have the nearly identical level of impact, while drought may not cause damage to infrastructure but results in a shortage of water.

The observations from Fig. 4 can be helpful for the authorities to prioritise the tasks that may need to be taken to restore the water supply in the post-disaster period and prepare plans to avert the damage to some extent.

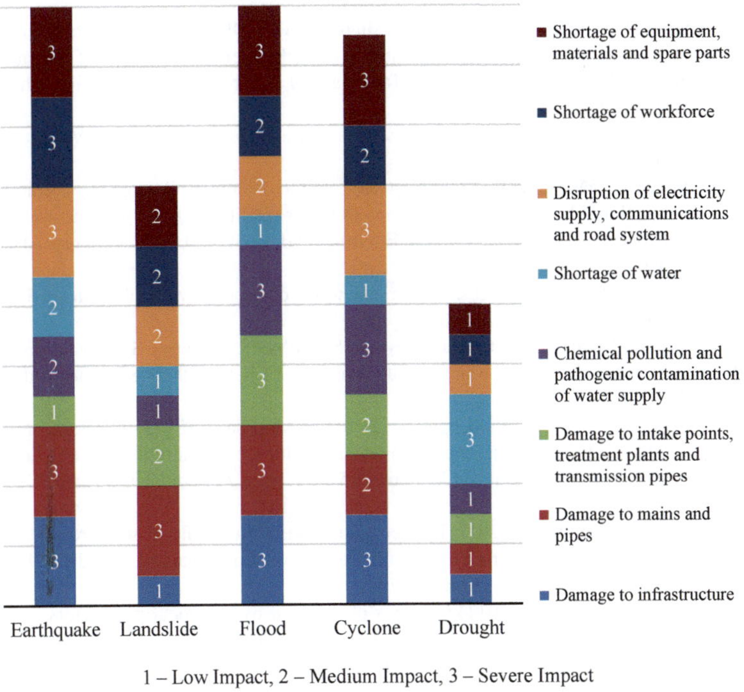

Fig. 4 Impact of disasters on water supply and sewerage system

4 Conclusion

With the changing climate and increasing human intervention, the dynamics of the balance of the world has shifted. It has resulted in an increased frequency of natural disasters. Different disasters create different scenarios. A detailed investigation into the environmental causes and impacts is vital in preparing for these disasters and reducing their effects. The effects on water sources help in identifying the essential parameters that needs our attention. The detailed analysis of disaster impacts can help us prepare better and create technologies and innovations that will be useful in dealing with adverse effects.

References

Adikari Y, Yoshitani J (2009) Global trends in water-related disasters: an insight for policymakers. United Nations World Water Assess. Program. Publ. pp 1–24

Barik SK, Muduli PR, Mohanty B, Behera AT, Mallick S, Das A, Samal RN, Rastogi G, Pattnaik AK (2017) Spatio-temporal variability and the impact of Phailin on water quality of Chilika lagoon. Cont Shelf Res 136:39–56. https://doi.org/10.1016/j.csr.2017.01.019

Chandrasekharan H, Sarangi A, Nagarajan M, Singh VP, Rao DUM, Stalin P, Natarajan K, Chandrasekaran B, Anbazhagan S (2008) Variability of soil-water quality due to Tsunami-2004 in the coastal belt of Nagapattinam district Tamilnadu. J Environ Manage 89:63–72. https://doi.org/10.1016/j.jenvman.2007.01.051

CRED (2020) EM-DAT The international disaster database. https://www.emdat.be/. Last Accessed 31 Dec 2020

Dhawde R, Surve N, Macaden R, Wennberg AC, Seifert-Dähnn I, Ghadge A, Birdi T (2018) Physicochemical and bacteriological analysis of water quality in drought prone areas of Pune and Satara districts of Maharashtra, India. Environ—MDPI 5:1–20. https://doi.org/10.3390/environments5050061

Grayman WM (2011) Water-related disasters: a review and commentary. Front Earth Sci 5:371–377. https://doi.org/10.1007/s11707-011-0205-y

Kamalanandhini M, Kalaivizhi R, Golda Percy VP, Srividhya S, Dheepak S, Thiyaneshwaran KK (2019) Effect of flood event on water quality. Rasayan J Chem 12:849–854. https://doi.org/10.31788/RJC.2019.1225232

Kamalanandhini M, Annadurai R, Dheepak S, Deepak P, Metilda JE (2021) Assessment of hydrological drought condition and its impact on water quality-a case study in parts of Chengalpattu district, Tamil Nadu, India. Rasayan J Chem 14:51–57. https://doi.org/10.31788/RJC.2021.1416095

Mishra A, Alnahit A, Campbell B (2021) Impact of land uses, drought, flood, wildfire, and cascading events on water quality and microbial communities: a review and analysis. J Hydrol 596:125707. https://doi.org/10.1016/j.jhydrol.2020.125707

Mitra A, Halder P, Banerjee K (2011) Changes of selected hydrological parameters in Hooghly estuary in response to a severe tropical cyclone (Aila) Indian. J Mar Sci 40:32–36

PAHO (2002) Emergencies and disasters in drinking water supply and sewerage systems : guidelines for effective response. Washington DC, USA

Saha A, Salim SM, Sudheesan D, Suresh VR, Nag SK, Panikkar P, Das BK (2020) Impacts of a massive flood event on the physico-chemistry and water quality of river Pampa in Western Ghats of India. Int J Environ Anal Chem 00:1–19. https://doi.org/10.1080/03067319.2020.1843026

Shen G, Hwang SN (2019) Spatial–Temporal snapshots of global natural disaster impacts revealed from EM-DAT for 1900–2015. Geomatics Nat Hazards Risk 10:912–934. https://doi.org/10.1080/19475705.2018.1552630

Singh VS (2008) Impact of the earthquake and Tsunami of December 26, 2004, on the groundwater regime at Neill Island (south Andaman). J Environ Manage 89:58–62. https://doi.org/10.1016/j.jenvman.2007.01.049

Srinivas H, Nakagawa Y (2008) Environmental implications for disaster preparedness: lessons Learnt from the Indian Ocean Tsunami. J Environ Manage 89:4–13. https://doi.org/10.1016/j.jenvman.2007.01.054

Violette, S, Boulicot G, Gorelick SM (2009) Tsunami-induced groundwater salinization in southeastern India. Comptes Rendus—Geosci 341:339–346. https://doi.org/10.1016/j.crte.2008.11.013

Young S, Balluz L, Malilay J (2004) Natural and technologic hazardous material releases during and after natural disasters: a review. Sci Total Environ 322:3–20. https://doi.org/10.1016/S0048-9697(03)00446-7

Modeling of Ground Water Quality Index of Haryana: A Comparison of Deep Neural Network and Support Vector Machine

Hemant Raheja, Arun Goel, and Mahesh Pal

1 Introduction

Groundwater is considered main source of water for domestic, municipal, industrial and irrigation throughout the world (Sefie et al. 2018; Zhang et al. 2017a). Water is covered about 75% surface area of the world and total quantity is about 1386 M km^3. Out of which 97.5% i.e. 1351.4 M km^3 of the earth's water lies in oceans as saline, which is not suitable for use by the humans. Only 2.5% i.e. 35 M km^3 is available as fresh water. Only 10.6 M km^3 is available as fresh water in reservoirs, rivers, lakes, streams, and ground water out of the total of 24.4 M km^3, which is frozen in the polar ice caps. At present, increasing pollution, climate change and human activity are serious problems, leading to deterioration of fresh water sources. This deterioration has an adverse effect on water consumption, leading to a shortage of drinking water. Several studies have identified various processes to measure changes in groundwater quality such as Hydro-chemical, Hydrodynamic, Multivariate statistical techniques, and Isotopic analysis. The key groundwater quality indicators are: pH, TH, Electrical conductivity (EC), Calcium (Ca^{2+}), Magnesium (Mg^{2+}), Sodium (Na^+), Potassium (K^+), Chloride (Cl^-), Sulfate ($SO_4{}^{2-}$), Bicarbonate ($HCO_3{}^-$), Nitrate ($NO_3{}^-$), bicarbonate ($HCO_3{}^-$), Carbonate ($CO_3{}^{2-}$), Fluoride (F^-), Silicon Dioxide (SiO_2).

In the last two decades, support vector machine (SVM) and neural networks (NNs) based regression process are designed to be a powerful tool for the application of water-resources engineering (Zhang et al. 2017b; Minasny and Perfect 2004; Pal

H. Raheja (✉) · A. Goel · M. Pal
Civil Engineering Department, NIT Kurukshetra, Haryana 136119, India
e-mail: raheja.hemant788@gmail.com

A. Goel
e-mail: darun_goel@nitkkr.ac.in

M. Pal
e-mail: mahesh.pal@nitkkr.ac.in

K. K. Singh and C. S. Prasad Ojha (eds.), *Sustainable Management of Land, Water and Pollution of Built-up Area*, Society of Earth Scientists Series,
https://doi.org/10.1007/978-3-031-56176-4_12

and Goel 2006; Elbisy 2015; Najah Ahmed et al. 2019; El Bilali et al. 2021; Raheja et al. 2021). In the last decade, deep neural network (DNN) has become a very useful in the field of civil engineering (Kumar and Abraham 2019; Dick et al. 2019; Ding et al. 2019; Nguyen et al. 2019; Pal 2019; Wu et al. 2018). In this study, water quality index has been calculated based on chemical parameters such as is pH, TH, electrical conductivity (EC), sodium (Na^+), magnesium (Mg^{2+}), calcium (Ca^{2+}), potassium (K^+), chloride (Cl^-), Sulphate ($SO_4{}^{2-}$), bicarbonate ($HCO_3{}^-$), nitrate ($NO_3{}^{2-}$), carbonate ($CO_3{}^{2-}$), fluoride (F^-) and silicon dioxide (SiO_2) by using DNN and SVM modelling techniques for the purpose of water quality in Haryana state.

2 Machine Learning Approaches

2.1 Support Vector Machines (SVM)

Statistical learning theory is derived by SVM and the classification problems can be solved by SVM which is based on optimal division of classes. For classification SVM, choose a classifier that has minimize the error from an infinite number and to minimize the structural risk, adjust a limit to error. The maximum margin between the two classes could be calculated from the selected hyperplane by adding the hyperplane's distance from the closest point to the two classes (Pal and Goel 2006).

2.2 Deep Neural Network (DNN)

An artificial neural network, or ANN, is a mathematical model for replicating a network of biological neurons that imitates the human brain and allows computers to learn and make decisions similarly to people. Deep learning, also known as DNN, has recently piqued the interest of both academics and industry professionals. Deep neural network (DNN) is powerful methods for solving large-scale real-world problems (Loc et al. 2020). DNN is a technique in which a huge number of nodes and two or more two hidden layers advanced mathematical modeling is used, which is commonly known as DNN.

3 Material and Methods

3.1 Measurement of Water Quality Index (WQI)

This method has been used to evaluate groundwater quality for the drinking purposes in different geological regions (Adimalla and Qian 2019; Adimalla et al. 2020; Gaikwad et al. 2020). The WQI calculations include four steps (Horton 1965).

1. Weight (wi) was assigned to fourteen parameters according to the relative importance of the overall drinking water quality. The weight 1 is for the least significant and 5 is for most significant parameters. In this study, the weight of 5 for $NO_3{}^{2-}$, F^-, SiO_2 due to its high importance in assessment of groundwater quality (Ramakrishnaiah et al. 2009). Ca^{2+}, Mg^{2+} and Na^+ have been given a weight of 2 because these are less toxic whereas the weight of 1 is for remaining parameters.
2. The relative weight (Wi) is calculated from as given below in Eq. (1)

$$W_i = \frac{wi}{\sum_{i=0}^{n} wi} \tag{1}$$

3. The scale (qi) is determined using Eq. (2), and the qi is given before being multiplied by 100. Then, by dividing it's WHO-provided concentration limits, the findings are obtained.

$$q_i = \frac{C_i}{S_i} \times 100 \tag{2}$$

where, S_i = acceptable limits of drinking water standard as per WHO and C_i = chemical parameter concentration.

4. The water quality index (WQI) is calculate from Eq. (3), firstly SI_i can be determined by multiply W_i and q_i.

$$WQI = \sum_{i=1}^{n} SI_i \tag{3}$$

Later on calculated values of WQI are divided in five categories. When the WQI > 300 represents the unsuitable for drinking purpose. When the WQI values lies in between the 200 to 300, then it represents a very poor water quality. When the WQI values in between the 100 to 200, then it represents poor water quality. When the WQI values in between the 50 to 100, then it represents that water quality is good and when values < 50 then quality of water is excellent and is suitable for drinking (Wang et al. 2017).

3.2 Data Set Used

The water samples were collected by the Central Ground Water Board (CGWB), North Western Region (NWR) Chandigarh from national groundwater monitoring stations distributed over 20 districts of Haryana during May 2015. A total of 309 observations were used in this study. 247 samples are used for the training and 62 samples for testing. The pH, Total hardness (TH), Electrical conductivity (EC), sodium (Na^+), potassium (K^+), calcium (Ca^{2+}), magnesium (Mg^{2+}), Sulphate ($SO_4{}^{2-}$), bicarbonate ($HCO_3{}^-$), nitrate ($NO_3{}^{2-}$), carbonate ($CO_3{}^{2-}$), chloride (Cl^-), fluoride (F^-), silicon dioxide (SiO_2) were taken as input parameter and the WQI was considered as output.

3.3 Performance Evaluation Indices

The computed values are assessed using the ensuing goodness-of-fit metrics:

- *Coefficient of correlation (CC)*: expressed in Eq. (4)

$$CC = \frac{m\left(\sum_{i=1}^{m} a_i b_i\right) - \left(\sum_{i=1}^{m} a_i\right)\left(\sum_{i=1}^{m} b_i\right)}{\sqrt{m \sum_{i=1}^{m}(a_i^2) - \left(\sum_{i=1}^{m} a_i\right)^2][m \sum_{i=1}^{m}\left(b_i^2\right) - \left(\sum_{i=1}^{m} b_i\right)^2]}} \tag{4}$$

- *Root mean square error* $(RMSE)$: expressed in Eq. (5)

$$RMSE = \sqrt{\frac{1}{m}(\sum_{i=1}^{m}(a_i - b_i)^2} \tag{5}$$

- *Mean absolute error* (MAE): expressed in Eq. (6)

$$MAE = \sum_{i=0}^{m}|a_i - b_i| \tag{6}$$

where, a_i = observed values, b_i = predicted values and m = number of observations.

Table 1 Final input parameter of SVM modeling technique

Approach	User-defined parameters
SVM	Kernel = PUK, omega (ω) = 10.0 sigma (σ) = 5.0 Epsilon parameter (ε) = 0.001

4 Results and Discussion

4.1 SVM

The performance of SVM may affect by the choice of suitable kernel function in the used datasets, results with three different kernels i.e. Radial Basis Function (RBF), Normalized Polynomial Kernel (NPKF), and Pearson Universal Kernel (PUK) functions were compared. Results in terms of CC value with test data suggest improved performance by PUK kernel function (0.9759), followed by the Normalized Polykernel (0.8799), radial basis function (RBF, 0.8506) with a used dataset. Table 1 show that the user defined parameter of SVM modeling technique.

4.2 DNN

In this study, the behavior of DNN with different values of hyperparameters, three hidden layers is found to be working well with the used dataset. Figures 1, 2, 3 and 4 shows the RMSE values variation with different methods of Weight Initialization, Number of batch sizes, Number of epochs, and Number of nodes in the hidden layers using testing data. Results have shown that three hidden layers having 100, 50 and 30 nodes with 400 epochs perform best with this dataset. Figure 1 shows the relation between the weight initialization and RMSE. From this figure, it is clear that Xavier weight initialization is performing best in terms of RMSE value (i.e., 1.8847). Figure 2 provides a plot between the Number of Epochs and RMSE. From this figure, it is clear that epochs = 400 are performing best as the value of RMSE is the lowest in this case. Figure 3 represents the relation between batch size and RMSE. It is clear from this figure that increasing the batch size beyond 3 increases root mean square error also increases, suggesting that a batch size of 3 is performing well with this data. Figure 4 provide a plot between the number of nodes in three hidden layer and RMSE. This figure clearly indicates that the hidden layer having 100, 50 and 30 nodes perform best with the used datasets. Table 2 show that the user defined parameter of DNN modeling technique.

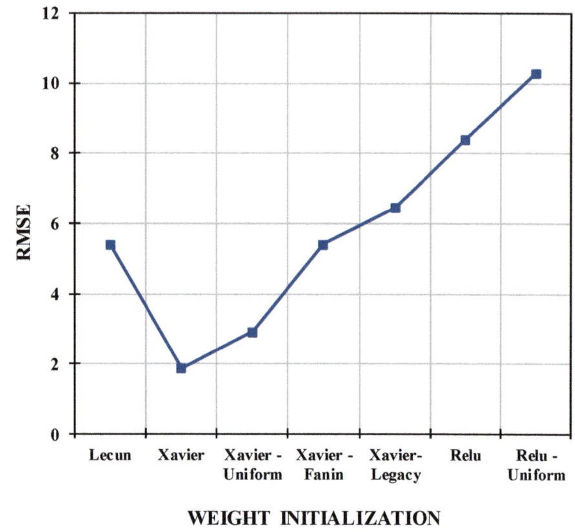

Fig. 1 Variation of RMSE with weight initialization

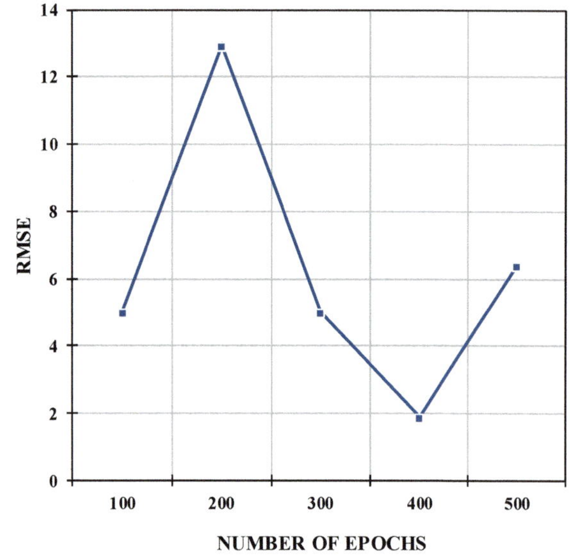

Fig. 2 Variation of RMSE with number of epochs

4.3 Comparison of SVM and DNN

A comparative analysis of predictive performances using SVM with three different kernel functions and DNN is provided in Table 3. Figure 5 shows a variation of actual and predicted WQI values by using DNN in testing phase. Figure 6 demonstrates the

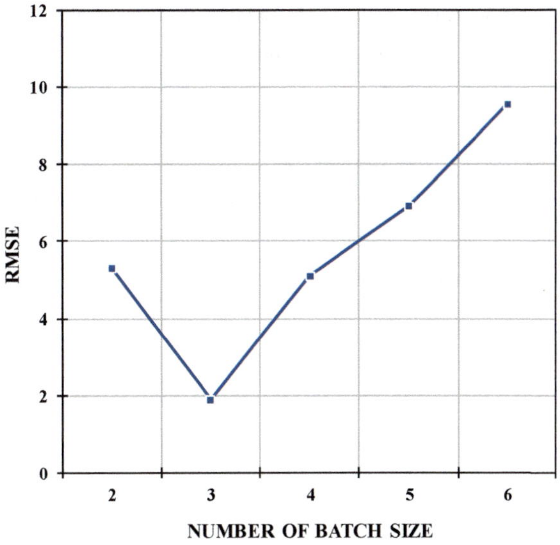

Fig. 3 Variation of RMSE with batch size

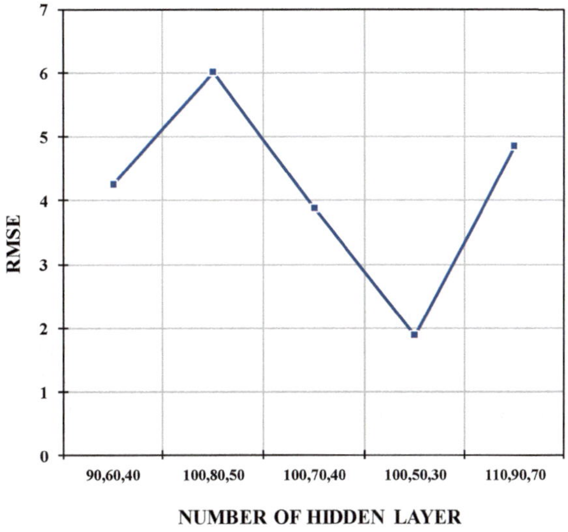

Fig. 4 Variation of RMSE with number of hidden layers

Table 2 Final input parameter of DNN modeling technique

Approach	User-defined parameters
Deep neural network	Hidden layers = 3 (100, 50, 30 nodes) Dropout layers = 2 with p = 0.3 Batch size = 3 Epochs = 400 Weight initiation = XAVIER Activation function = ReLU

Table 3 Values of CC, MAE, and RMSE values with a test dataset

Sr. No.	Name of algorithm	CC	MAE	RMSE
1	DNN	0.9999	1.336	1.8847
2	SVM with normalized polykernel	0.8799	33.59	86.39
3	SVM with PUK	0.9759	7.91	44.46
4	SVM with RBF kernel	0.8506	19.21	96.07

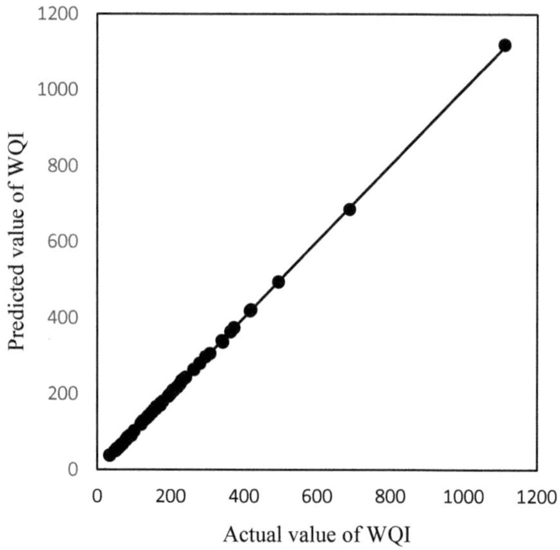

Fig. 5 Comparison of actual versus predicted values WQI using DNN in testing phase

variation of actual and predicted WQI values based on SVM using PUK kernel function in the testing phase. From the Figs. 5 and 6 suggests that the DNN is performing better than the SVM because prediction of DNN shows near the agreement line. Figure 7 signifies the actual and predicted WQI using DNN and SVM algorithm in testing phase.

5 Conclusions

Analytical study has been attempted to compare the DNN and SVM (Radial Basis function (RBF), Normalized Polynomial Kernel (NPKF), and Pearson Universal Kernel (PUK) functions) modeling techniques in predicting water quality index of groundwater in the Haryana state (India). The main conclusion from this study is the superior performance of the DNN approach for water quality index in comparison to extensively used SVM. Other important conclusions from this study are that in spite of improved predictive performance by DNN (CC = 0.9999), it is affected

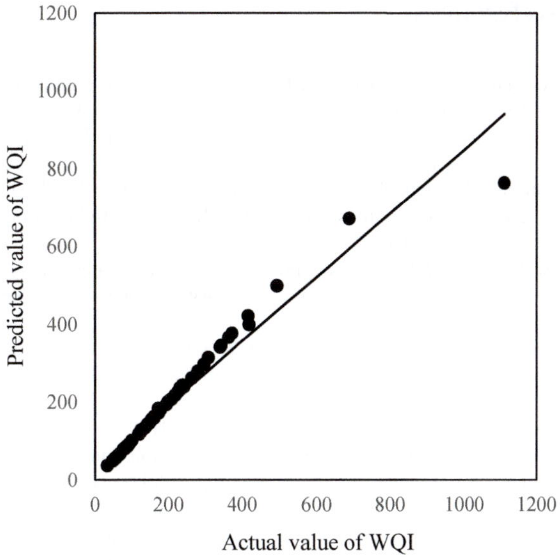

Fig. 6 Comparison of actual versus predicted values of WQI using SVM with PUK Function in testing phase

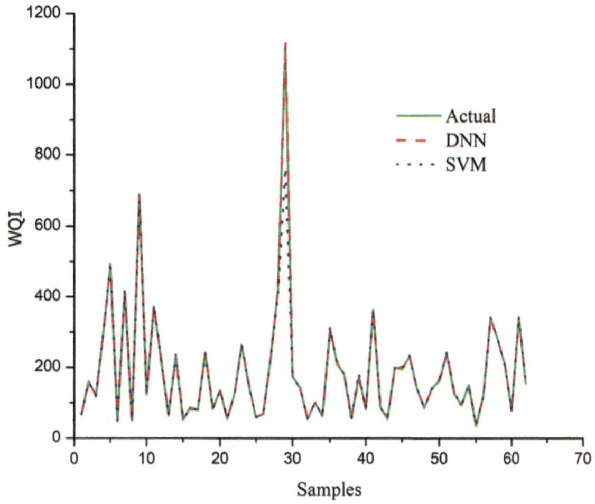

Fig. 7 Actual and predicted WQI using DNN and SVM algorithm in testing phase

by the choice of user-defined parameters for the datasets. This suggests the need of obtaining suitable values of various user-defined parameters for the datasets at hand. In the case of SVM, out of three kernel functions, PUK kernel function (CC = 0.9759) has performed best in comparisons to other two kernel function.

Acknowledgements Hemant Raheja, the first author, expresses gratitude to the CGWB (Chandigarh) for supplying the data as well as to the MHRD, GOI for funding the current work through a Ph.D. scholarship award (2K19/NITK/PHD/61900011).

References

Adimalla N, Qian H (2019) Groundwater quality evaluation using water quality index (WQI) for drinking purposes and human health risk (HHR) assessment in an agricultural region of Nanganur, south India. Ecotoxicol Environ Saf 176(126):153–161. https://doi.org/10.1016/j.ecoenv.2019.03.066

Adimalla N, Dhakate R, Kasarla A, Taloor AK (2020) Appraisal of groundwater quality for drinking and irrigation purposes in Central Telangana, India. Groundw Sustain Dev 10(126):100334. https://doi.org/10.1016/j.gsd.2020.100334

El Bilali A, Taleb A, Brouziyne Y (2021) Groundwater quality forecasting using machine learning algorithms for irrigation purposes. Agric Water Manag 245:106625. https://doi.org/10.1016/j.agwat.2020.106625

Dick K, Russell L, Souley Dosso Y, Kwamena F, Green JR (2019) Deep learning for critical infrastructure resilience. J Infrastruct Syst 25(2):05019003. https://doi.org/10.1061/(ASCE)IS.1943-555X.0000477

Ding F, Zhang Z, Zhou Y, Chen X, Ran B (2019) Large-scale full-coverage traffic speed estimation under extreme traffic conditions using a big data and deep learning approach: case study in China. J Transp Eng Part A Syst. https://doi.org/10.1061/jtepbs.0000230

Elbisy MS (2015) Support vector machine and regression analysis to predict the field hydraulic conductivity of sandy soil. KSCE J Civ Eng. https://doi.org/10.1007/s12205-015-0210-x

Gaikwad SK et al (2020) Assessment of the groundwater geochemistry from a part of west coast of India using statistical methods and water quality index. Hydro Res. https://doi.org/10.1016/j.hydres.2020.04.001

Horton RK (1965) An index number system for rating water quality. J Water Pollut Control Fed 37(3):300–306

Kumar SS, Abraham DM (2019) A deep learning based automated structural defect detection system for sewer pipelines. https://doi.org/10.1061/9780784482445.029

Loc HH, Do QH, Cokro AA, Irvine KN (2020) Deep neural network analyses of water quality time series associated with water sensitive urban design (WSUD) features. J Appl Water Eng Res 8(4):313–332. https://doi.org/10.1080/23249676.2020.1831976

Minasny B, Perfect E (2004) Solute adsorption and transport parameters. Dev Soil Sci. https://doi.org/10.1016/S0166-2481(04)30012-7

Najah Ahmed A et al. (2019) Machine learning methods for better water quality prediction. J Hydrol 578. https://doi.org/10.1016/j.jhydrol.2019.124084

Nguyen T, Kashani A, Ngo T, Bordas S (2019) Deep neural network with high-order neuron for the prediction of foamed concrete strength. Comput Civ Infrastruct Eng. https://doi.org/10.1111/mice.12422

Pal M (2019) Deep neural network based pier scour modeling. ISH J. Hydraul. Eng. 00(00):1–6. https://doi.org/10.1080/09715010.2019.1679673

Pal M, Goel A (2006) Prediction of the end-depth ratio and discharge in semi-circular and circular shaped channels using support vector machines. Flow Meas Instrum 17(1):49–57. https://doi.org/10.1016/j.flowmeasinst.2005.05.003

Raheja H, Goel A, Pal M (2021) Prediction of groundwater quality indices using machine learning algorithms. Water Pract Technol 00:1–16. https://doi.org/10.2166/wpt.2021.120

Ramakrishnaiah CR, Sadashivaiah C, Ranganna G (2009) Assessment of water quality index for the groundwater in Tumkur taluk, Karnataka state, India. E-J Chem 6(2):523–530. https://doi.org/10.1155/2009/757424

Sefie A et al (2018) Hydrogeochemistry and groundwater quality assessment of the multilayered aquifer in Lower Kelantan Basin, Kelantan, Malaysia. Environ Earth Sci 77(10):1–15. https://doi.org/10.1007/s12665-018-7561-9

Wang X, Zhang F, Ding J (2017) Evaluation of water quality based on a machine learning algorithm and water quality index for the Ebinur Lake Watershed, China. Sci Rep 7(1):1–18. https://doi.org/10.1038/s41598-017-12853-y

Wu Z, Wang X, Chen Y, Cai Y, Deng J (2018) Assessing river water quality using water quality index in Lake Taihu Basin, China. Sci Total Environ 612:914–922. https://doi.org/10.1016/j.scitotenv.2017.08.293

Zhang X, Miao J, Hu BX, Liu H, Zhang H, Ma Z (2017a) Hydrogeochemical characterization and groundwater quality assessment in intruded coastal brine aquifers (Laizhou Bay, China). Environ Sci Pollut Res 24(26):21073–21090. https://doi.org/10.1007/s11356-017-9641-x

Zhang Z, Wang Y, Chen P, Yu G (2018) Application of long short-term memory neural network for multi-step travel time forecasting on urban expressways. In: CICTP 2017, January 2018, pp 444–454. https://doi.org/10.1061/9780784480915.046

Sustainable Pollution Management

Removal of Priority Pollutants Using Photocatalysis and Wastewater Remediation

Sandeep Singla, Rishav Garg, and Rajni Garg

1 Introduction

Addressing the issue of non-biodegradable contaminants in wastewater is a major focus in many countries (Eddy et al. 2022). Economic efficiency is calculated by the elimination of all sorts of harmful chemicals. Industrial effluents or wastewater containing numerous contaminants must be treated in accordance with international standards (Ajmal et al. 2014). Many pollutants need high priority treatment to achieve water quality standards and recommendations to reduce discharge, because such pollutants are commonly classified as priority pollutants in wastewater. Priority contaminants are the significant pollutants, including heavy metals and certain organic chemicals that pose negative effect on environment and population (Gómez-Pastora et al. 2014). The "Clean Water Act" (USA) has designated some of the toxic pollutants as "priority pollutants" in their guidelines for water quality and effluent treatment. These pollutants have been given high priority since these are commonly found in wastewater discharged from various industries (Singh et al. 2022; Garg et al. 2020a). The objective of the Priority Pollutants Database is to provide a useful and handy tool for scientists, researchers, and analysts for study, monitoring, analysis and control of the chemicals that can cause occupational, physiological and environmental issues (Garg et al. 2020b). It should be noted that any such pollutant that is insoluble in water or is present in total dissolved form, is capable

S. Singla (✉)
Chitkara University Institute of Engineering and Technology, Chitkara University, Punjab, India
e-mail: sandeep.singla@chitkara.edu.in

R. Garg
Department of Civil Engineering, Galgotias College of Engineering and Technology, Greater Noida, India

R. Garg
R&D Department, Institute of Sci-Tech Affairs, Mohali, India

K. K. Singh and C. S. Prasad Ojha (eds.), *Sustainable Management of Land, Water and Pollution of Built-up Area*, Society of Earth Scientists Series,
https://doi.org/10.1007/978-3-031-56176-4_13

of contaminating significant volumes of water (Naseem et al. 2018). Thus, even a small amount of such pollutants in aquatic resources can be dangerous. Further, the common water and industrial treatment processes result in the formation of intermediates or by-products that can themselves lead to problems (Xu et al. 2012).

It must be ensured that any form of effluent discharged into waterways meet the accepted allowable thresholds (Garg et al. 2022a). The amount of pollutants allowed to be discharged are directly proportionate to the type of pollutant detected in the discharged effluent. Hence, the appropriate and suitable treatment process must result in the compliance of these guidelines (Bashambu et al. 2021). Some methods of removing or degrading organic compounds include biological treatment, chemical oxidation, air desorption, liquid–liquid separation, adsorption, reverse osmosis and ultra-filtration etc. (Naseem and Durrani 2021). Based on the composition of the wastewater or industrial effluent, the destructive degradation procedures can be used including chemical oxidation, incineration, and decay, which would certainly allow the effective elimination of the pollutant in the aqueous form; and non-destructive procedures, such as liquid–liquid extraction and absorption, which would permit the recovery of the pollutant (Dong et al. 2014). The use of one or the other process will depend on the amount and nature of the chemical for each process. It is important to select the most suitable approach in order to select the most effective wastewater treatment method. It is obvious that different environmental technologies are needed according to location and concentration of pollutants (Singh et al. 2019). For example, for low concentration of pollutants, chemical oxidation, membrane use and adsorption process can be effectively used. But for the large concentration of pollutants, incineration as well as chemical oxidation techniques can be used. However, in current scenario, the most promising approach is photochemical as the advance oxidation process (AOP) (Munoz et al. 2015).

It is important to completely degrade the priority contaminants but is not possible via treatment methods that are established and have been proven effective for other primary pollutants (Vanthana Sree et al. 2020). In contrast, destructive oxidation process can be used to transform organic pollutants into simple and non-toxic species including carbon dioxide, water, and/or oxidized anions (Al-Hussain et al. 2018). Such advanced oxidation processes are collectively referred to as AOPs. The process takes place through the production of free radicals that are extremely reactive. These processes also require the addition of certain oxidants and exposure to light radiations or suitable catalysts (Bashambu et al. 2021). Nano chemistry works at the interface of science and technology with diverse applications in various fields including water remediation, construction sector and pharmaceuticals. Literature reports use of numerous nanomaterials for instance nano-silica (Garg et al. 2022b), nano-zinc oxide (Sharma et al. 2020), nano-titania (Ajmal et al. 2014) etc. The researchers select semiconductors as photocatalysts because of the narrow energy gap between valence and conduction bands (Garg and Garg 2020). To continue with photocatalysis, semiconductors need to consume energy greater than or at least equal to this energy gap. Electron–hole migration forms electronically charged electron/hole pairs, resulting in positively charged holes (Garg et al. 2022c). The holes could lead to degradation of the donor molecules as illustrated in Fig. 1. The use of TiO_2 as a catalyst in

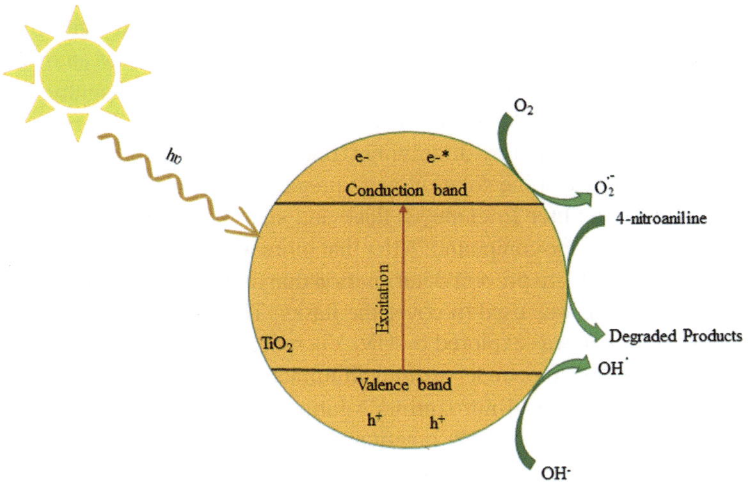

Fig. 1 Mechanism for photodegradation of pollutants by nanoparticles

AOPs have been well documented in the literature (Ajmal et al. 2014). The effluent treatment can be done by using photocatalytic oxidation. Among potential semiconductors, TiO_2 is most commonly used because it has many advantages (Albukhari et al. 2019). This is due to its inertness and resistance towards corrosion in addition to the advantage that it can be easily and economically processed after fabrication. Under moderate operating conditions, this will be useful. The greatest drawback is the need for ultraviolet light for photocatalysis (Pi et al. 2018).

The study has been planned to reduce the amount of pollution by using the photocatalysis. 4-nitroaniline was selected as a typical compound being a persistent and non-biodegradable compound that can't be degraded by conventional treatment techniques. The focus of the study was the photocatalytic oxidation of 4-nitroaniline as a priority pollutant in industrial effluents. Literature survey was carried out to determine available lists of substances assigned as priority pollutant by Environmental Protection Agency. It was preferred to conduct the degradation studies on 4-nitroaniline because of their scarce scientific literature. The studies in terms of absorbance shift, effect of concentration of catalyst, pH, light intensity, concentration of pollutant and COD etc. have been used to reveal the information about suitability of the proposed system for treatment of industrial effluents.

2 Materials and Methods

Double distilled water was used to prepare all solutions. 4-nitroaniline, Degussa P-25 TiO_2 and other chemicals were purchased from Merck Pvt. Ltd. and were used as such without any further purification. The UV–Vis spectrum of 4-nitroaniline was taken with the assistance of ultraviolet–visible spectrophotometer and maximum absorbance was observed at 325 nm. Hydrogen peroxide was used as an oxidant. pH of the solution was modified using HCl and NaOH solutions. Test solution concentration was selected before the oxidation tests could be carried out. The concentration of 30 mg/L was formulated by dissolving 0.030 g of the compound in one liter of distilled water. 4-nitroaniline solution were prepared by adding known quantity of the compound to one-liter Erlenmeyer flask and small amount of deionized water was added to dissolve the compound. After that more water was added to fill the flask up to the mark. In order to prevent deterioration due to fluorescent lights in the laboratory, aluminum foil was used to cover the flasks. The samples were analyzed for COD and degradation was explored by UV–Vis spectrophotometer analysis. Photocatalytic studies were done under various illumination conditions of UV radiations as well as solar radiations. 4-nitroaniline solution was subjected for photocatalytic reaction with variation of various reaction parameters including concentration of catalyst and sample, pH and the reaction parameters were finally optimized. After photocatalytic treatment of the wastewater (optimized conditions), the sample was purified and analyzed for color, pH, and COD (Dong et al. 2014).

3 Results and Discussion

Photocatalytic oxidation method was used to degrade 4-nitroaniline solution in presence of titanium oxide catalyst. A model was constructed by randomizing the variables, which include dosage of catalyst, compound and oxidant, pH, UV exposure time and solar light exposure time for each experiment. 4-nitroaniline was exposed to photocatalytic treatment under different conditions and the effectiveness of the treatment was assessed (Vanthana Sree et al. 2020). The two parameter, area and volume also affect the rate of oxidation of contaminants. If the volume decreases and the area increases, more of light will come into contact with the solution, and reaction rate increases i.e., more degradation will be resulted due to increase in surface area (Bashambu et al. 2021). Reverse happens on decreasing the area and increasing the volume. As a result, as container surface area increased, deterioration increased as well. Further, the influence of ultraviolet radiation, solar radiation, and intensity in addition to variation of pH and dosage of catalyst, oxidant and compound have been discussed ahead.

3.1 UV–Vis Spectral Analysis

The catalytic addition in presence of light radiations has a significant impact on the degradation rate of 4-nitroaniline. After UV irradiation, 99% of dyes had been depleted within 270 min. In the slurry mode, the Degussa P-25 TiO_2 was used as catalyst to study the effect of solar and UV radiations on photocatalytic degradation

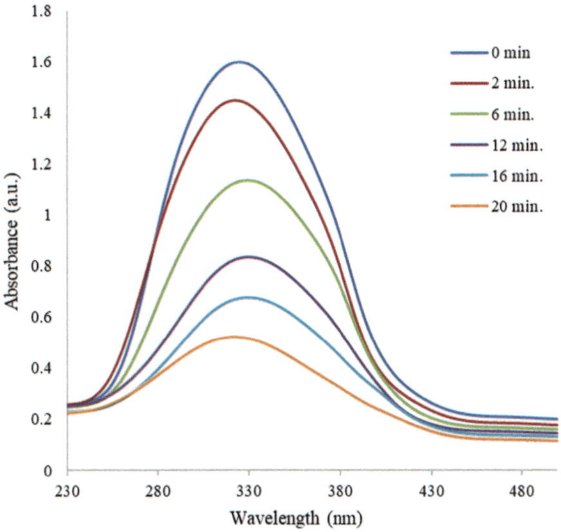

Fig. 2 UV spectra of 4-nitroaniline with change in time

of the compound. Analysis was done in both the absence and presence of the nano-catalyst in the presence of ultraviolet and solar light. The color of sample solution has been observed to become light in color indicating the degradation of the compounds that was depicted as change in intensity of absorbance spectra (Pi et al. 2018). It was found that in presence of UV radiations and solar radiations, 65% and 34% degradation of the compound was observed in absence of the catalyst, respectively.

Figure 2 displays the absorption spectra of p-nitroaniline with absorption maxima at 325 nm with change in time. The variation in peak intensity at 325 nm for all of the solutions was used to estimate the rate of deterioration (Dong et al. 2014). During the analysis, the addition of catalyst resulted in an increase in degradation (Munoz et al. 2015). Thus, in presence of UV radiations and solar radiations, 93% and 62% degradation of the compound was observed in presence of the catalyst, respectively. The positive effects have been observed in conjunction with catalyst in both UV and solar radiations. Catalytic degradation increased significantly due to the absorption of more photons due to catalysts (Pi et al. 2018). Researchers have also studied the chemical adsorption process of aniline oxidation on TiO_2. The simulation study indicated that the aromatic ring is oriented flat and the titanium ion is trapped in a plane of the surface (Al-Hussain et al. 2018).

3.2 *Impact of pH*

The pH of the 4-nitroaniline sample solutions was determined before irradiation. The pH is important because it affects the production of hydroxyl radicals that are an effective oxidizing agent for the photocatalytic reactions (Albukhari et al. 2019).

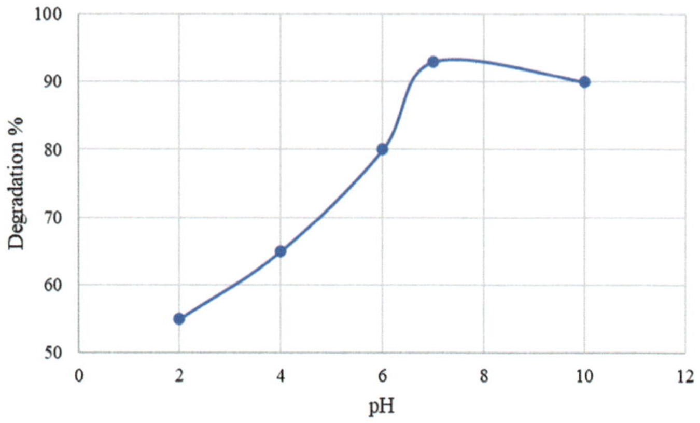

Fig. 3 Impact of pH on degradation

Nitroaniline oxidation reaction was studied at low pH by using Fenton and electro-Fenton process and the kinetics was investigated by many scientists and it has been found that addition of catalysts increases the degradation of these photocatalytic, electro-Fenton processes/photoelectron-Fenton methods (Munoz et al. 2015). The used dosage of catalyst was 2 g/L for all the experiments and the pH of the samples (25 mg/L) was varied from 2–10. It was observed that as the pH of the system was increased from 2–10, the amount of degradation increased (Fig. 3). The sample was found to degrade faster as the pH was increased till pH 7 with slight decrease afterwards and a maximum degradation of 93% at pH 7 was observed. After the photo-catalytic treatment of sample, the pH excellent for biological treatment as well as discharge of wastewater into waterbodies can be attained (Pi et al. 2018). Based on the results obtained, the pH 7 was picked to be the best pH to achieve photodegradation.

3.3 *Impact of Catalyst Dosage*

The second important parameter to the degradation rate is the catalytic dosage, that strongly affects the kinetics for degradation of nitroaniline (Naseem et al. 2018). The factors of pH and TiO_2 dosage have been observed to have an effect on the rate of degradation. Literature reports photocatalytic degradation of various dyes under titania samples in aqueous medium (Ajmal et al. 2014). Researchers have also used titania catalyst to study the rate of degradation of ethidium bromide and acridine orange. The increase in the degradation was found with increase in catalytic dosage up to an optimum concentration and the later concentration of the catalyst makes the degradation occur slower (Dong et al. 2014).

In the slurry mode, the Degussa P-25 TiO_2 was used as catalyst for the photocatalytic oxidation. The optimum catalytic concentration was determined by varying the

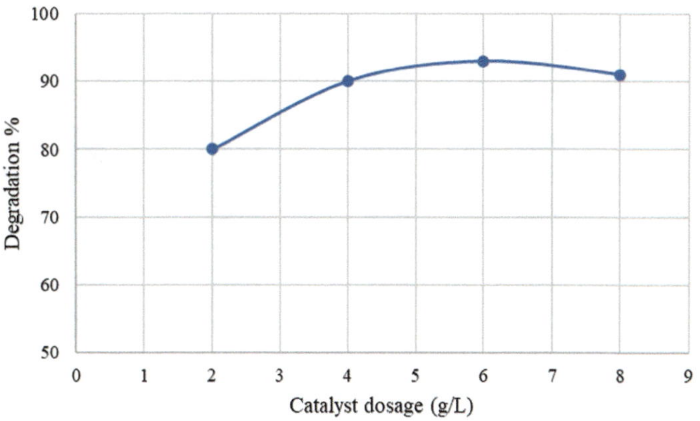

Fig. 4 Impact of catalyst dosage

catalysts dosage from 2–8 g/L at the optimum pH (7) for photocatalytic oxidation of nitroaniline solutions (25 mg/L). As the concentration of catalyst was increased from 2–8 g/L, the percentage of degradation was found to vary significantly. The degradation was found to first increase till the catalyst dosage up to 6 g/L and then was found to decrease (Fig. 4). The initial increase in catalytic degradation with increase in catalyst dosage can be ascribed to the increased absorption of more photons and more compounds of nitroaniline on the surface of catalyst that accelerate the reaction (Vanthana Sree et al. 2020). But adding more catalyst to the solution does not continuously increase the reactivity. It is due to the reason that all the reactant species already got absorbed. Further decrease in degradation at the increased dosage of catalyst can be attributed to the agglomeration of catalyst molecules and screening effect resulting in the decreased number of active sites and decreased degradation of compounds (Vanthana Sree et al. 2020). Thus, a catalyst dosage of 6 g/L for analyses of the photodegradation of 4-nitroaniline solution were found to be optimum dosage.

3.4 Impact of Compound Concentration

The concentration of pollutant is a critical parameter for degradation. After the optimization of the pH and catalytic dose, the impact of the concentration of 4-nitroaniline on its photocatalytic degradation was also investigated. The concertation of the compound was varied from 25 to 100 mg/L, at constant pH (7) and catalyst dosage of 6 g/L. It was found that with the increasing pollutant concentration from 25 to 100 mg/L, rate of degrading got lowered significantly (Fig. 5). A possible reason for this result is because of an increase in the concentration of compound, the level of adsorption on catalytic surface got increased, which decreased the catalytic efficiency of titanium dioxide (Vanthana Sree et al. 2020). Increased concentration of 4-nitroaniline solution also diminishes the path length of photon that enters in the

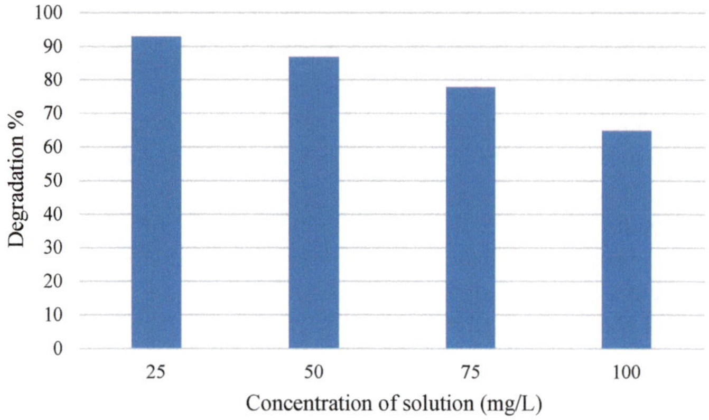

Fig. 5 Impact of compound concentration

sample solution. A large amount of ultraviolet light can be absorbed by the nitroaniline molecules, which may decrease the ability of the catalyst to catalyze the reaction (Pi et al. 2018).

COD is considered as an important factor for every pollutant that can be used to determine the extent of degradation of pollutant. The COD analysis of samples was found to decrease significantly both in presence of solar radiations and UV radiations respectively. The rate of photodegradation in presence of solar light radiations were found to be much higher as compared to artificial illumination. Photocatalytic oxidation and degradation of phenolic solutions has also been examined with natural solar light radiations and artificial illumination using titanium dioxide dissolved in a solvent or adsorbed on silica gel as a photo-catalyst (Rani et al. 2020; Zhao et al. 2019).

4 Conclusion

Photocatalysis is an innovative tool for photodegradation of toxic chemicals and the formation of minerals from organic pollutants as explored in here with the following conclusions:

1. 4-nitroaniline compound has been effectively degraded exploring TiO_2 as a photocatalyst.
2. The optimal dosing of TiO_2 was 6 g/L with a working pH of 7 for photodegradation of 4-nitroaniline.
3. According to the research on the sunlight, it is clear that solar light can be used for the oxidation and discoloration of the compound solution.
4. The change in the intensity and color of the spectra for 4-nitroaniline compound solutions indicates the conversion of compounds to simpler products.

5. There is also a strong correlation between the color, pH and concertation of the compound.
6. The concentration of COD in treated sample showed that in the case of UV and solar photocatalysis, concentration of COD is reduced effectively.

From the above results, it can be discovered that photochemistry can be used appropriately and effectively to degrade priority pollutants. Heterogeneous photocatalysis is an environmentally safe way to manage treatment of wastewater and to reduce the number of pollutants. The implications from the study are that both solar and UV light can be used for treatment of wastewater containing priority pollutants and can help in reducing their running costs.

References

Ajmal A, Majeed I, Malik RN, Idriss H, Nadeem MA (2014) Principles and mechanisms of photocatalytic dye degradation on TiO_2 based photocatalysts: a comparative overview. RSC Adv 4:37003–37026. https://doi.org/10.1039/C4RA06658H

Albukhari SM, Ismail M, Akhtar K, Danish EY (2019) Catalytic reduction of nitrophenols and dyes using silver nanoparticles @ cellulose polymer paper for the resolution of waste water treatment challenges. Colloids Surf A Physicochem Eng Asp 577:548–561. https://doi.org/10.1016/j.colsurfa.2019.05.058

Al-Hussain SA, Ezzat AO, Gaffer AK, Atta AM (2018) Removal of organic water pollutant using magnetite nanomaterials embedded with ionic copolymers of 2-acrylamido-2-methylpropane sodium sulfonate cryogels. Polym Int 67:166–177. https://doi.org/10.1002/pi.5492

Bashambu L, Singh R, Verma J (2021) Metal/metal oxide nanocomposite membranes for water purification. Mater Today Proc 44:538–545. https://doi.org/10.1016/j.matpr.2020.10.213

Dong Z, Le X, Li X, Zhang W, Dong C, Ma J (2014) Silver nanoparticles immobilized on fibrous nano-silica as highly efficient and recyclable heterogeneous catalyst for reduction of 4-nitrophenol and 2-nitroaniline. Appl Catal B Environ 158–159:129–135. https://doi.org/10.1016/j.apcatb.2014.04.015

Eddy NO, Garg R, Garg R, Aikoye AO, Ita BI (2022) Waste to resource recovery: mesoporous adsorbent from orange peel for the removal of trypan blue dye from aqueous solution. Biomass Convers Biorefinery. https://doi.org/10.1007/s13399-022-02571-5

Garg R, Garg R, Thakur A, Arif SM (2020a) Water remediation using biosorbent obtained from agricultural and fruit waste. Mater Today Proc 46:6669–6672. https://doi.org/10.1016/j.matpr.2021.04.132

Garg R, Kumari M, Kumar M, Dhiman S, Garg R (2020b) Green synthesis of calcium carbonate nanoparticles using waste fruit peel extract. Mater Today Proc 46:6665–6668. https://doi.org/10.1016/j.matpr.2021.04.124

Garg R, Garg R, Okon Eddy N, Ibrahim Almohana A, Fahad Almojil S, Amir Khan M, Ho Hong S (2022b) Biosynthesized silica-based zinc oxide nanocomposites for the sequestration of heavy metal ions from aqueous solutions. J King Saud Univ—Sci 34:101996. https://doi.org/10.1016/j.jksus.2022.101996

Garg R, Rani P, Garg R, Amir M, Ahmad N, Husain A, Heloisa J (2022c) Biomedical and catalytic applications of agri-based biosynthesized silver nanoparticles. Environ Pollut 310:119830. https://doi.org/10.1016/j.envpol.2022.119830

Garg R, Garg R (2020) Effect of zinc oxide nanoparticles on mechanical properties of silica fume-based cement composites. In: Materials today: proceedings. pp 778–783. https://doi.org/10.1016/j.matpr.2020.06.168

Garg RR, Garg RR, Khan MA, Bansal M, Garg VK (2022) Utilization of biosynthesized silica-supported iron oxide nanocomposites for the adsorptive removal of heavy metal ions from aqueous solutions. Environ Sci Pollut Res 1–10. https://doi.org/10.21203/rs.3.rs-1394501/v1

Gómez-Pastora J, Bringas E, Ortiz I (2014) Recent progress and future challenges on the use of high performance magnetic nano-adsorbents in environmental applications. Chem Eng J 256:187–204. https://doi.org/10.1016/j.cej.2014.06.119

Munoz M, de Pedro ZM, Casas JA, Rodriguez JJ (2015) Preparation of magnetite-based catalysts and their application in heterogeneous Fenton oxidation—a review. Appl Catal B Environ 176–177:249–265. https://doi.org/10.1016/j.apcatb.2015.04.003

Naseem T, Durrani T (2021) The role of some important metal oxide nanoparticles for wastewater and antibacterial applications: a review. Environ Chem Ecotoxicol 3:59–75. https://doi.org/10.1016/j.enceco.2020.12.001

Naseem K, Farooqi ZH, Begum R, Irfan A (2018) Removal of Congo red dye from aqueous medium by its catalytic reduction using sodium borohydride in the presence of various inorganic nano-catalysts: a review. J Clean Prod 187:296–307. https://doi.org/10.1016/j.jclepro.2018.03.209

Pi Y, Li X, Xia Q, Wu J, Li Y, Xiao J, Li Z (2018) Adsorptive and photocatalytic removal of persistent organic pollutants (POPs) in water by metal-organic frameworks (MOFs). Chem Eng J 337:351–371. https://doi.org/10.1016/j.cej.2017.12.092

Rani L, Kaushal J, Srivastav AL, Mahajan P (2020) A critical review on recent developments in MOF adsorbents for the elimination of toxic heavy metals from aqueous solutions. Environ Sci Pollut Res 27(36):44771–44796. https://doi.org/10.1007/s11356-020-10738-8

Sharma R, Garg R, Kumari A (2020) A review on biogenic synthesis, applications and toxicity aspects of zinc oxide nanoparticles. https://doi.org/10.17179/excli2020-2842

Singh S, Kumar V, Romero R, Sharma K, Singh J (2019) Applications of nanoparticles in wastewater treatment. In: Nanotechnology in the life sciences. pp 395–418. https://doi.org/10.1007/978-3-030-17061-5_17

Singh R, Pareek N, Kumar R, Vivekanand V (2022) Anaerobic biodigesters for human waste treatment anaerobic biodigesters for the treatment of high-strength wastewater. Springer, Singapore, pp 75–94

Vanthana Sree G, Nagaraaj P, Kalanidhi K, Aswathy CA, Rajasekaran P (2020) Calcium oxide a sustainable photocatalyst derived from eggshell for efficient photo-degradation of organic pollutants. J Clean Prod 270:122294. https://doi.org/10.1016/j.jclepro.2020.122294

Xu P, Zeng GM, Huang DL, Feng CL, Hu S, Zhao MH, Lai C, Wei Z, Huang C, Xie GX, Liu ZF (2012) Use of iron oxide nanomaterials in wastewater treatment: a review. Sci Total Environ 424:1–10. https://doi.org/10.1016/j.scitotenv.2012.02.023

Zhao Y, Liu L, Li C, Ye B, Xiong J, Shi X (2019) Immobilization of polyethyleneimine-templated silver nanoparticles onto filter paper for catalytic applications. Colloids Surf A Physicochem Eng Asp 571:44–49. https://doi.org/10.1016/j.colsurfa.2019.03.075

Mitigating Landfill Gas Emissions from Open Dumpsites Using Bio-Cover

Deval Singh, Anil Kumar Dikshit, and Sunil Kumar

1 Introduction

In most developing countries, the urban population has increased at a varying rate of 34% to 56% per year (CPCB 1999-2000). India is a nation with a rapid growth in population, modernization, and industrialization; with these factors ultimately leading to migration (of more than 31.2%) of rural habitat to urban cities (Census 2011). However, this transition in urban population has improved the nation's economic aspects, but at the cost of failure in the existing urban infrastructure like roads, housing, water supply, sanitation, solid waste management, various other amenities etc. (Pedersen and Bukh 2019). Studies have predicted that India might generate more than 440 X 10^3 tons/day (TPD) of MSW by 2026 (Pedersen and Bukh 2019). Typically, 75% MSW, which consist of biodegradable waste, plastic, paper, and others (textile, glass, metal, drain slit, street sweeping, and inert matter) with a percentage ratio of 51%, 10%, 7%, and 32%, respectively, is being dumped directly to the dumpsites (CPCB 1999-2000). The Planning Commission Report has claimed that ~ 377 million population residing in urban cities have an MSW generation potential of 60 million tons/year and this might surpass 160 million tons/year by 2031 (Planning Commission Report 2014). So, to take care of this quantity of MSW, India might require more than 40,000 ha of dumping area. In a country like India, there is a lack of actual primary data on per capita MSW generation, its characteristics, and impacts on the local environment (CPCB 2021). Therefore, it is hard to determine the precise land area requirement for MSW disposal. The survey, carried

D. Singh (✉)
Indian Institute of Technology Bombay, Powai, Mumbai 400076, India
e-mail: devsiingh99@gmail.com

A. K. Dikshit · S. Kumar
CSIR-National Environmental Engineering Research Institute (CSIR-NEERI), Nehru Marg, Nagpur 440020, India

K. K. Singh and C. S. Prasad Ojha (eds.), *Sustainable Management of Land, Water and Pollution of Built-up Area*, Society of Earth Scientists Series,
https://doi.org/10.1007/978-3-031-56176-4_14

out in 2004–2005 by the CSIR- NEERI in more than 50 cities, revealed that around 20 cities could generate 35,000 TPD of MSW in the upcoming few decades. Later, 59 cities were again surveyed by CPCB in 2009–2010 and the study concluded that the same 20 cities could generate more than 50,000 TPD of MSW (CPCB 2021). Similar trend is being observed all across the world as well. This change in the MSW generation pattern has motivated researchers across the globe to find alternative methods for MSW management and disposal.

Methane (CH_4) and carbon dioxide (CO_2) are the primary Landfill Gases (LFG) released from MSW dumpsites (Pedersen and Bukh 2019; Scheutz et al. 2009). The overall LFGs such as CH_4, CO_2, ammonia, hydrogen sulfide, and water vapor are ranked among the Greenhouse Gases (GHG) contributing toward global warming. The studies find that the impact associated with CH_4 emissions are more significant in comparison to other LFGs. The continuous emission of these LFGs may cause a change in atmospheric stability resulting in flagging of aerosol cooling and an increase in atmospheric temperature (Whalen et al. 1990).

The continuous non-scientific dumping of MSW has caused the piling up of old, partially decomposed, and fresh waste. This pilling of MSW leads to a decline inflow of oxygen at the bottom of the dumpsites, resulting in anaerobic conditions. Later, this leads to LFG emissions from the bottom of the dumpsite. The sudden exposure to CH_4 with the normal ambient condition may result in the auto-ignition of inorganic waste (plastic, paper, etc.) present at the dumpsite. This phenomenon of auto-ignition may further result in landfill fire.

Bio-cover is a capping layer specially designed using soil, clay, wood, organic compost, etc., to avoid direct emission of LFGs. Based on the past studies, the concept of bio-cover was found to be feasible for both small and larger dumpsites depending upon their area stretch (Abushammala et al. 2012). The physico-chemical parameters such as pH, moisture content, organic content, etc., can be managed using field operation; this control mechanism allows oxidization of CH_4 through bio-cover. The present study is focused on the application of bio-cover on open dumpsites for CH_4 mitigation. The impact of pH, temperature, oxygen concentration, soil texture, properties of bio-cover material, moisture, etc., on bio-cover design has also been discussed.

2 Kinetics of CH_4 Oxidation in Bio-Cover

The primary objective of bio-cover is to oxidize CH_4 by using oxidation layers (made of soil, organic compost, dry leaves, etc.) and gas dispersion layer (made of gravel and brick) (Huber-Humer 2004). Figure 1 shows typical bio-cover, where the gas dispersion layer acts as a basic foundation for the oxidation layer, as it allows LFGs to flow through the permeable layer. The oxidation layer is less porous and less permeable compared to the gas dispersion layer, which ultimately helps in the oxidization of CH_4 (Abushammala et al. 2012). The oxidative process facilitated by bio-cover is governed by Eq. 1:

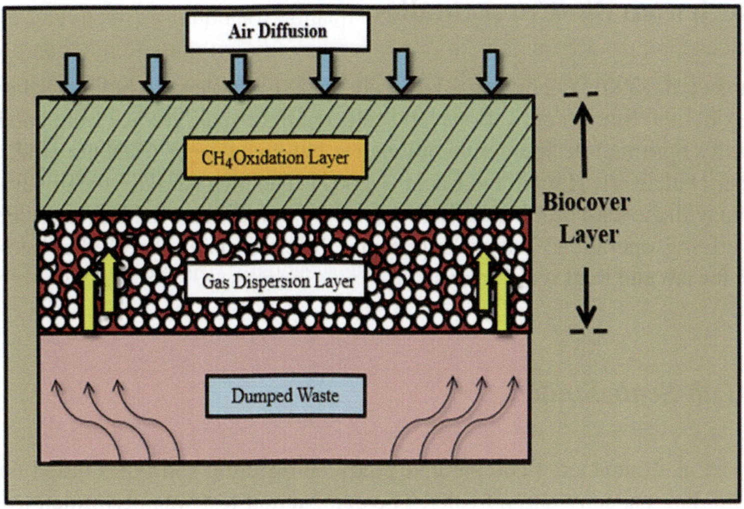

Fig. 1 Kinematic behavior of CH4 oxidation from bio-cover layer

$$CH_4 + 2O_2 \rightarrow CO_2 + H_2O + \text{microbial biomass} \quad (1)$$

The height and depth of the dumpsite in most of the developing countries may vary from 5 to 15 m depending upon the age of waste. The top, middle, and bottom surface of the dumpsite comprises fresh (1 to 2 days), partial (5 to 12 months), and old MSW (1 to 10 years). The studies have proved that the top layer of mixed waste and soil residue comprises naturally occurring methanogens, potent enough to oxidize CH4 in a limiting proportion (Abushammala et al. 2012). However, after a certain time frame, the decline in oxygen and nutrients level in the top surface layer of dumpsite results in the decline of methanogens. Once appropriate bio-cover is provided, it becomes easier to maintain the level of oxygen and nutrients in the top surface, resulting in long-term oxidation of LFGs released from the dumpsite. The bio-cover spread layer acts as a CH4 oxidizing media to convert LFGs into CO2, water, and waste biomass (Albanna et al. 2007). Researchers across the globe have demonstrated the application of bio-cover made from locally available material such as soil, wood, tire chips, dried leaves, etc. (Huber-Humer 2004; Kettunen et al. 2006; Eun-Hee-Lee et al. 2014; Rachor et al. 2011). The methanotrophs present in these materials are aerobic, which are efficient enough to oxidize CH4.

3 Review on Bio-Cover Studies

Albanna et al. (2007) conducted a lab-scale study to implement bio-cover layer for CH4 oxidation. Since then, various lab-scale and field-scale studies were conducted to classify optimum design combinations for bio-cover layer (Scheutz et al. 2009; Visscher et al. 2001; Humer and Lechner 2001). The idea is to use locally available material with the least processing facility (sieving, grinding, shredding, etc.) required for overlying operation. Table 1 represents different materials used in bio-cover design for lab and field scale analysis.

3.1 Lab-Scale Studies

Rachor et al. conducted a comparative study on different soil types based on their water holding capacity and porosity. The compaction test was performed to determine the water seepage capacity in soil bio-cover. A lab-scale study was conducted using three soil columns filled with 80 cm thickness and 17 cm gas dispersion layer made from gravel layer. The soil bio-cover was compacted at three proctor densities of 75%, 80%, and 85%, respectively. Moisture content for samples was customized to the field limit. The final study determined a more substantial relationship between oxygen and soil bio-cover. The results concluded that 75% proctor density could completely oxidize CH_4 influx of about 3.5 g CH_4/m^2/day. The preliminary studies comparing soil particle, texture, surface area, and organic content were also characterized (Rachor et al. 2011). The conclusive evidence was reported that mineral soil samples were more compatible with bio-cover layer (He et al. 2011).

Eun-Hee et al. conducted a lab-scale study on bio-cover to identify the impact of methanotrophic microbes in CH4 oxidation. The soil samples were collected from the top (0–10 cm), center (10–40 cm), and base (40–50 cm) layers. The study proved that a higher percentage of active methanogen was found to be at a depth of 5 to 30 cm, respectively (Eun-Hee-Lee et al. 2014). A similar study was conducted by He et al. to identify the change in parameters such as microbes, pH, temperature, moisture content, and oxygen concentration of bio-cover in China. This study determined that methanogenic microbes are sensitive toward moisture content in the soil strata, which also gets affected by oxygen absorption (He et al. 2011). It was also identified that type II methanotrophs were most predominant in CH_4 oxidation (Zhang et al. 2014).

Zhang et al. studied the impact of ammonium (NH^+) concentrations of 100, 300, 600, and 1200 mg/kg on the oxidation process. The results of the study suggested that a permissible limit of 100 to 300 mg/kg of NH^+ concentration can be used as a standard benchmark for optimum CH4 oxidation. The balanced proportion of NH^+ allows enhancing growth rate in methanogenic microbes resulting in a higher oxidation rate (Zhang et al. 2014). Changgen et al. carried out the study to design bio-cover layer using green waste aging not more than 24 months. The batch scale column study was carried out for 15 months to estimate the CH4 oxidation rate from

Table 1 Different materials used in bio-cover design

Author	Material	Type of study	CH_4 oxidation efficiency (%)	Description
Eun-Hee et al 2014)	Soil mixture	Lab-scale	Middle layer = 85 Bottom layer = 71	The column study was performed to analyse the maximum oxidation rate at varying depth of bio-cover
Rachor et al. (2011)	Sands, silts, gravel, clay	Lab-scale	90–100	The study was carried to determine the impact of soil compaction on bio-cover layer
Changgen et al. (2015)	Fresh and aged green waste	Lab-scale		The use of green waste aging more than 24 months was monitored and analyzed for more than a year, to determine the oxidation efficiency
Chiemchaisri et al. (2013)	Plastic and organic waste	Field-scale	Stabilizes waste layer up to 34.1 g/m^3d	The CH_4 generation was studied with respect to seasonal variations. Study concluded higher amount of CH_4 generation in summer season
Scheutz et al. (2017)	Compost (75% grass clippings and 25% cow manure)	Field-scale	81- 100	Field scale study on MSW landfill site was performed by using compost as a bio-cover oxidation material
Lee et al. (2014)	Sandy soil, vegetation, NH4	Field-scale	60	Ammonia was used as influencing reagent to improvise the CH_4. Oxidation rate
Stern et al. (2007)	Sandy clay, fine sandy loam, thick layer of glass cullet, thick mulch layer	Lab-scale		Study suggested that methanotrophic bacteria's and temperature variation can be considered as most influencing factors for CH_4 oxidation
Gebert and Groengroeft (2006)	Gravel, soil	Lab-scale	Layer 1 = 75 Layer 2 = 85 Layer 3 = 95	The proctor density test was performed. It was conclusive that soil bio-cover with 75% Proctor density had highest CH_4 oxidation (3.5 g CH_4/m^2/day)
Scheutz et al. (2014)	Local compost material	Field-scale	80	Bio-cover comprised of both organic and inorganic materials

fresh and old green waste. The final research concluded that oxidation rate varies from 140 to 200 gm/m^2/day for fresh and old waste, respectively. The study also concluded that oxidation rate is highly influenced by seasonal variation, resulting in the decline of oxidation rate in the winter season (Changgen et al. 2015).

3.2 *Field-Scale Studies*

Chanton et al. carried out the study to report CH_4 concentration and oxidation efficiency of bio-cover layer on active landfill site. The preliminary study was conducted to identify hotspot zone for CH4 emission from landfills sites. The hotspot zones were marked from both plane and inclined geology of landfill. The concentration of CH_4 emissions from the plane and inclined landfill sites were reported. The bio-cover was prepared from locally available soil. The oxidation efficiency was monitored for a year; the idea was to analyze the impact of different seasonal variations on bio-cover efficiency. The study reported a higher CH4 oxidation rate in the summer season (Chanton et al. 2011).

Chiemchaisri et al. used old and dried MSW as a bio-cover layer. The study relieved that plastic and dried organic waste fractions were the primary components to promote CH4 oxidation. The loosened texture and tiered fraction of plastic waste allow easy penetration of the atmospheric oxygen from the top substrate of bio-cover. At the same time, the organic fraction maintains the required proportion of moisture content and nutrient inside the bio-cover (Chiemchaisri et al. 2013).

Scheutz et al. conducted a study on unlined landfill sites in Denmark. The locally available compost was used as a bio-cover layer; this resulted in 80% CH_4 oxidation efficiency (Scheutz et al. 2014). Pawłowska et al. also conducted a similar study using a piolet scale system for the same landfill site with a partial submissive bio-cover layer of a local compact. The oxidation efficiency varied from 80% to 100%. Both the study demonstrated conclusive evidence to prove the implementation of compost as bio-cover. The compost has a physical property to maintain its inner heat of hydration; this allows effective CH_4 oxidation in the winter season. Apart from this, compost is a cost-effective material readily available near landfill sites (Pawłowska and Stępniewski 2008).

Bomin et al. conducted a study using a pilot-scale set up at a South Korean landfill site. The landfill site was enclosed using soil bio-cover. The pilot-scale setup was designed to monitor CH_4 oxidation rate at a depth of 10, 15, 30, and 50 cm. The overall efficiency of 70% was reported in the month of February and May. However, the optimum CH_4 oxidation rate was achieved at a depth interval of 15 to 30 cm (Bomin et al. 2016).

4 Factors Influencing Bio-Cover

4.1 Soil Texture

Gebert et al. carried out a study on different soil samples to determine soil compatibility for bio-cover. The result of the study suggested that coarse-grained soil is most compatible for landfill bio-cover. The porosity of soil plays a vital role in CH_4 dissemination from the surface and subsurface layer of bio-cover (Gebert et al. 2011). A similar study was conducted by Pawłowska and Stępniewski to determine the effective soil particle size required for bio-cover. The study revealed that soil particle ranging from 0.5 to 1 mm has a higher oxidation efficiency (Pawłowska and Stępniewski 2008).

4.2 Organic Content

The studies have proved that organic content is an influencing parameter for maintaining the microbes' ratio inside the bio-cover. Christophersen et al. conducted a study to determine the role of the organic fraction in oxidation efficiency. The study recommended using manure, biomass, yard waste, and compost with a C/N ratio ranging from 25 to 35. The balanced proportion of carbon and nitrogen allows microbial cell formation and maintains the food by biomass ratio (Christophersen et al. 2000).

4.3 Moisture Content

The studies have recommended limiting proportion (40% to 50%) of moisture content for enhanced oxidation results. It was observed that a decline in oxidation rate is influenced by excess moisture content as it restricts the microbial regeneration process inside the bio-cover. In some cases, extra moisture content causes a decline in oxygen penetration; this results in a decrease in the biological metabolism of methanotrophic microbes (Visscher et al. 2001).

4.4 Temperature

Temperature is one of the most influencing factors for microbial growth rate in bio-cover (Visscher et al. 2001). Mor et al. conducted a comparative study to understand the interdependence of temperature variation on microbial growth rate and oxidation efficiency. The study results recommended that temperature variation ranging from

15 to 30 °C is the optimum range for bio-cover design (Mor et al. 2006). De Visscher et al. conducted similar research suggesting 35 °C as the most favorable temperature for the oxidation process (Visscher et al. 2001). The study also suggested the interdependence of temperature variation and moisture content on microbial growth rate (Tecle et al. 2008). Castaldi and Fierro also recommended moisture content range 30% to 40% and temperature range 35–45 °C (Castaldi and Fierro 2005).

4.5 PH

The high pH variation has been observed in the bio-cover layer depending upon its life span. The range varying from 6 to 8 is highly conducive for microbial growth (Pol et al. 2007). The study states that oxidation is affected depending upon fluctuation in pH values. According to Saari et al. the optimum pH value should range between 4 and 7.5 (Saari et al. 2004), while Hanson et al. has concluded that a pH value of less than 5 can cause a decline in the oxidation process. However, the pH variation may differ depending upon material type and its chemical properties (Hanson and Hanson 1996).

4.6 Oxygen Concentration

A balance concentration of oxygen is required for aerobic microbes to carry out the process of oxidation in bio-cover layer. Studies have suggested that oxygen concentration should be in lower proportion when compared with the atmospheric level (Abushammala et al. 2014). In most cases, oxygen concentration varies depending upon the depth of the bio-cover layer, soil type, soil texture, pH, metrological factors, and moisture content (Lee et al. 2014).

5 Summary and Recommendation

The open dumpsites are considered to be the primary source of CH_4 emission. Recent and past studies have performed lab-scale and field-scale analyses to develop the CH_4 oxidation layer in the form of bio-cover. The results of studies have suggested that locally available materials such as organic compost, green waste, organic waste, and different soil grades have enough potential to oxidize CH_4 from open dumpsites. The use of methanogenic microbes to impact CH_4 oxidation is the primary weapon used in designing bio-cover. The influencing factors like soil texture, organic content, pH, temperature, etc., must be examined before implementing bio-cover layer. Based on the existing literature study, it found that the use of inorganic waste fraction is irrelevant in bio-cover design. In a developing country like India, no such scientific method of landfilling has been practiced. It has been observed that open dumpsites of solid waste with no capping provision have led to health hazardous, landfill fire, and change in air quality standards and increase in atmospheric temperature. As of now, no such practical application was found to be reported in the Indian context, but the idea to implement locally available material in bio-cover design can be easily implemented on a small pilot scale at existing dumpsites.

References

Abushammala M, Basri N, Basri H, Kadhum A, ElShafie A (2012) Empirical gas emission and oxidation measurement at cover soilof dumping site: example from Malaysia. Environ Monit Assess 185:4919

Abushammala M, Noor E, Dani I, Mohammad K (2014) Methane oxidation in landfill cover soils: a review. Asian J Atmos Environ 8:1

Albanna M, Fernandes L, Warith M (2007) Methane oxidation in landfill cover soil; the com- bined effects of moisture content, nutrient addition, and cover thickness. J Environ Engg Sci 6:191

Bomin K, Ryu G, Jeon J, Cho K (2016) Mitigation of methane and odor using a pilot-scale engineered bio-cover at a landfill in South Korea. American Geophysical Union, Fall General Assembly

Castaldi S, Fierro A (2005) Soil atmosphere methane exchange in undisturbed and burned mediterranean shrubland of Southern Italy. Ecosyst 8:182

Census (2011) Provisional population totals, India, http://censusindia.gov.in/2011-prov-results/datafiles/india/povpoputotalpresentation2011.pdf. Last Accessed 23 Aug 2021

Changgen M, Ramin Y, Byunghyun H, Erfan M, Jean V, Paul L (2015) Performance of green waste bio-covers for enhancing methane oxidation. Waste Manag 39:205

Chanton J, Abichou T, Langford C, Spokas K, Hater G, Green R (2011) Observations on the methane oxidation capacity of landfill soils. Waste Manage 31:914

Chavan D, Kumar S (2018) Reduction of methane emission from landfill using biocover as a biomitigation system: a review. Indian J Exp Biol 56:451–459

Chiemchaisri W, Chiemchaisri C, Boonchaiyuttasak J (2013) Utilization of stabilized wastes for reducing methane emission from municipal solid waste disposal. Bioresour Technol 141:199

Christophersen M, Linderod L, Jensen P, Kjeldsen P (2000) Methane oxidation at low tem- peratures in soil exposed to landfill gas. J Environ Qual 29:1989

CPCB (1999–2000) Status of solid waste generation, collection, treatment and disposal in metro cities (Series: CUPS/46/1999–2000), CPCB, New Delhi

CPCB (2021) Status of municipal solid waste generation, collection, treatment and dis- posal in class I cities, CPCB, New Delhi (Series: ADSORBS/31/1999–2000). http://www.bvucoepune.edu.in/pdf's/Research%20and%20Publication/Research%20Publication_200708/International%20Conference_2007-08/Municipal%20solid%20Prof%20MR%20Gidde.pdf. Last accessed on 23 Aug 2021

De Visscher A, Schippers M, Cleemput O (2001) Short-term kinetic response of enhanced methane oxidation in landfill cover soils to environmental factors. Biol Fertil Soils 33:231

Eun-Hee-Lee K, Tae G, Kyung S (2014) Depth profiles of methane oxidation potentials and methanotrophic community in a lab-scale bio-cover. J Biotechnol 184:56

Gebert J, Grongroft A (2006) Passive landfill gas emission Influence of atmospheric pressure and implications for the operation of methane-oxidising biofilters. Waste Manage 26:245

Gebert J, Groengroeft A, Pfeiffer E (2011) Relevance of soil physical properties for the micro- bial oxidation of methane in landfill covers. Soil Biol Biochem 43:1759

Hanson R, Hanson T (1996) Methanotrophic bacteria. Microbiol Rev 60:439

He P, Yang N, Fang W, Lu F, Shao L (2011) Interaction and independence on methane oxi- dation of landfill cover soil among three impact factors: water, oxygen and ammonium. Front Environ Sci Eng China 5:175

Huber-Humer M (2004) International research into landfills gas emissions and mitigation strategies-IWWG working group "CLEAR." Waste Manag 24:425

Humer M, Lechner P (2001) Design of a landfill cover layer to enhance methane oxidation; results from a two-year field investigation. In: Proceeding of SARDINIA 2001-eighth international waste management and landfill symposium, Cagliari, vol 1. pp 541

Kettunen R, Einola J, Rintala J (2006) Landfill methane oxidation in organic soil columns at low temperature. Water Air Soil Pollut 177:313

Lee S, Im J, Dispirito A, Bodrossy L, Barcelona M, Semrau J (2014) Effect of nutrient and selective inhibitor amendments on methane oxidation, nitrous oxide production, and key gene presence and expression in landfill cover soils. Environ Biotechnol 8:1

Mor S, De Visscher A, Ravindra K, Dahiya R, Chandra A, Cleemput O (2006) Induction of enhanced methane oxidation in compost: temperature and moisture response. Waste Manage 26:381

Pawłowska M, Stępniewski W (2008) An influence of methane concentration on the methanotrophic activity of a model landfill cover. Ecol Eng 26:392

Pedersen A, Bukh G (2019) Processes in a compost-based landfill bio-cover; methane emission, transport and oxidation, Denmark Technical University (DTU) library

Planning Commission Report (2014) Reports of the task force on waste to energy (Vol-I) (in the context of Integrated MSW management). Retrieved from http://planningcommis-sion.nic.in/reports/genrep/rep_wte1205.pdf. Last Accessed 23 Aug 2021

Pol A, Heijmans K, Harhangi H, Tedesco D, Jetten M, Opdencamp H (2007) Methanotrophy below pH1 by a new Verrucomicrobia species. Nat 450:874

Rachor I, Gebert J, Grongroft A, Pfeiffer E (2011) Assessment of the methane oxidation capacity of compacted soils intended for use as landfill cover materials. Waste Manag 31:833

Saari A, Rinnan R, Martikainen P (2004) Methane oxidation in boreal forest soils: kinetics and sensitivity to pH and ammonium. Soil Biol Biochem 36:1037

Scheutz C, Kjeldsen P, Bogner JE, Visscher AD, Gebert J, Hilger HA (2009) Microbial methane oxidation processes and technologies for mitigation of landfill gas emissions. Waste Manag 27:409–455

Scheutz C, Rasmus B, Per H, Jorgen H, Inmaculada M, Jacob G, Samuelsson J, Kjeldsen P (2014) Mitigation of methane emission from an old unlined landfill in Klintholm, Denmark using a passive bio-cover system. Waste Manag 34:1179

Scheutz C, Filippo C, De S, Kjeldsen P (2017) Mitigation of methane emissions in a pilot-scale bio-cover system at the AV Miljo Landfill, Denmark: 2 methane oxidation. Waste Manag 63:203

Stern J, Chanton J, Abichou T, Powelson D, Yuan L, Escoriza S, Bogner J (2007) Use of biologically active cover to reduce landfill methane emissions and enhance methane oxidation. Waste Manag 27:1248

Tecle D, Lee J, Hasan S (2008) Quantitative analysis of physical and geotechnical factors af- fecting methane emission in municipal solid waste landfill. Environ Geol 56:1135

Whalen S, Reeburgh W, Sandbeck K (1990) Rapid methane oxidation in a landfill cover soil. Appl Environ Microbiol 56:3405

Zhang X, YanKong J, Fang F, Xia Yao S, Ruo H (2014) Effects of ammonium on the activity and community of methanotrophs in landfill bio-cover soils. Syst Appl Microbiol 37:296

Bioreactor Landfills: Sustainable Solution to Waste Management

Lagudu S. Avinash and Anumita Mishra

1 Introduction

The rapid growth of population, industrialization and economic growth with high consumption of goods all over the world has led to the generation of 2.01 billion tons of municipal solid waste (MSW) annually, which is expected to increase by 70% by 2050 (The World Bank 2022; Kaza et al. 2050). In many countries, landfilling is the most economical way for disposing the MSW (Alkaabi et al. 2009; Simões and Catapreta 2013). The generation of waste is mainly from industrial, households, commercial enterprises, agricultural etc. The composition of global waste is shown in Fig. 1a (Kaza et al. 2050). The rules of waste management are based on the principles of sustainable development. These principles are made mandatory in municipalities to habituate the people to act in responsible manner environmentally, accountable and restoring balance under umbrella law of Environment Protection Act, 1986 (Environment and protection act: An umbrella legislation for environmental related laws (1986)). In many countries, waste management operations are of local responsibility and 70% of countries established institutions for regulatory oversight and policy developments. The treatment and disposing of globally generated MSW is shown in Fig. 1b. Food and greens are the largest wastes, which contribute to about 44% of global waste. In case of controlled landfills, there is no landfill gas management system which creates disturbance to environment and public health. Landfills are more advantageous than other disposing facilities because of the benefits associated with it (Somani et al. 2019).

L. S. Avinash (✉) · A. Mishra
Indian Institute of Technology Roorkee, Roorkee 247667, India
e-mail: ls_avinash@ce.iitr.ac.in

A. Mishra
e-mail: anumita.mishra@ce.iitr.ac.in

K. K. Singh and C. S. Prasad Ojha (eds.), *Sustainable Management of Land, Water and Pollution of Built-up Area*, Society of Earth Scientists Series,
https://doi.org/10.1007/978-3-031-56176-4_15

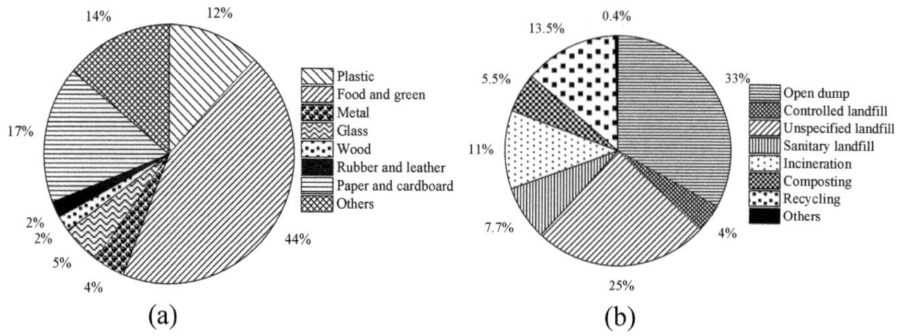

Fig. 1 **a** Composition of global waste **b** Treatment and disposal of global waste

A bioreactor landfill is a sanitary landfill that operates by sequential recirculation of liquids and air through the waste for enhancing the microbial process. The favorable conditions for micro-organisms are maintained to accelerate the rate of decomposition through breakdown of biodegradable fraction for accomplishing the stabilization (Reinhart et al. 2002; Kumar et al. 2011; Berge et al. 2005). The operation of bioreactor landfill is a sustainable solution to waste management since it minimizes the emission of greenhouse gases, stops the contamination of water bodies due to leachate percolation and reduces long-term monitoring (Warith 2002; Mali Sandip et al. 2012). In addition to liquid addition, other strategies such as shredding of waste, waste composition and age, adjustment of pH, addition of nutrients, management of temperature optimize the process of bioreactor landfill (Aromolaran and Sartaj 2021). Bioreactor landfill technology generates revenue from its end products to produce electricity and residues of mining yields various products such as bio soil, refuse derived fuel (RDF), daily covers for landfills as shown in Fig. 2 (Chembukavu et al. 2019; Nanda and Berruti 2021). With the advancement of technology for processing, treatment and disposal of waste, the yield products from waste act as a crucial component which favors in declining the volume of refuse from disposal. It also avails in transforming refuse into renewable resource and organic manure.

This study is intended to raise the awareness on bioreactor landfills which has a good potential for efficient management of MSW. The sustainable solution to waste management is achieved by proper utilization of the end-products such as leachate, landfill gas and landfill mining from bioreactor landfills. With this view, a literature review has been carried out related to the above mentioned issues and has been presented in this study.

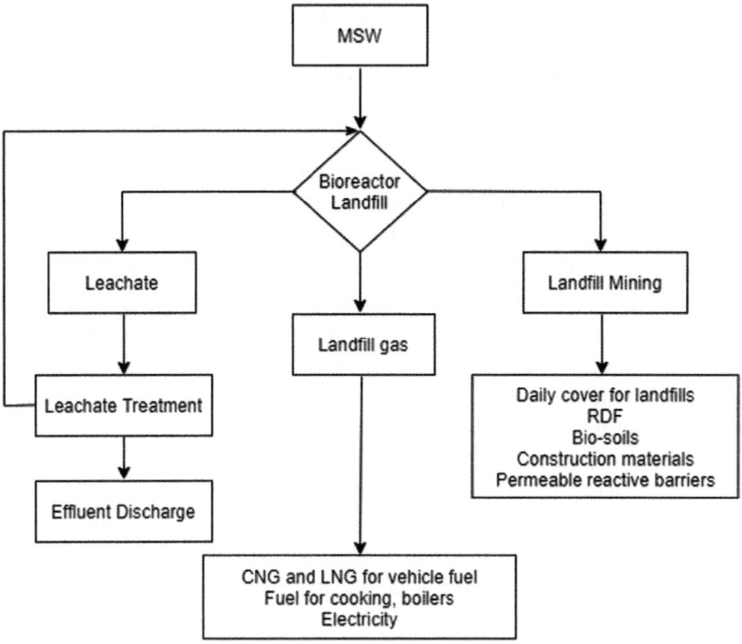

Fig. 2 Sustainable development of MSW through Bioreactor landfills

2 Sustainable Development Strategies

The focus of this study is to implement practices that maximize resource recovery and energy from waste and reduce the greenhouse gas emissions. Leachate can cause contamination of surface water and groundwater. Landfill gas emissions has the ability to absorb heat and prevent it escaping from into the atmosphere, which leads to global climate change. Landfill mining yields valuable products through reuse and recycling the waste. In this section, a discussion has been presented on the utilization of end-products of bioreactor landfills to achieve the goals of sustainable solid waste management. The literatures have been reviewed and presented for the following three aspects:

1. Landfill Leachate
2. Landfill gas
3. Landfill mining

2.1 Landfill Leachate

Leachate is a soluble product generated due to decomposed organic matter and infiltration of liquid percolating through waste layers (Reinhart et al. 2002). Leachate management has been recognized as one of the greatest problem of waste management because the leachate can pollute the surrounding water bodies, soils, and groundwater. Landfills in developing countries lack installation of engineered facilities such as liners and leachate collection systems (Ayub 2019). Landfill leachate consists of dissolved and suspended constituents. The acceleration of landfill stabilization is based on the characteristics of leachate (Suna Erses and Onay 2003). Young leachate has low pH, more organic fraction during the acid phase while the old leachate has high pH, readily reduced biodegradable organic fraction under methanogenic phase (Tchobanoglous et al. 1993). The effluent from leachate treatment is used for leachate recirculation or simply discharged in water bodies.

Leachate recirculation is the process of injecting and collecting the liquid from the bottom of landfill is recycled and again introduced into the landfill waste to maintain the moisture content (Al-kaabi 2007). The concept of liquid addition to enhance the degradation process through moisturizing the waste was proposed in 1970's (Pohland 1980). Influents into landfills are taken from various sources such as storm water runoff, landfill leachate, surface water, ponds, ground water, saline water, wastewater effluents from commercial and industries (Chembukavu et al. 2019; Ayub 2019). Liquids must be supplemented to waste to maintain optimum moisture content of 35–65% for accelerating the degradation mechanism. The moisture content in conventional landfills is 20–30% which is less than the optimum range (Guérin et al. 2004; Townsend 1998).

The sludge from different waste treatment plants and effluents from industries can be injected into bioreactor landfills. Therefore, it provides a sustainable solution for environmentally safe and economic disposal of waste products. It implies that these by-products that are perceived as wastes in one perspective can act as a valuable resource in another perspective. The various types of biosolids and their advantages of utilization in bioreactor landfills are shown in Table 1.

Leachate injection is carried out with a combination of liquid and municipal wastewater sludge at different rates in bioreactor to optimize the degradation of waste. They observed that increased injection rates of leachate and sludge, decreases the waste stabilization from 47 to 27 weeks (Rendra et al. 2007). Aromolaran and Sartaj (Aromolaran and Sartaj 2021) studied the injection of different blended leachate to enhance the methanogenesis process of MSW in bioreactor landfills and observed cumulative gas production is 2.5 times more compared to leachate recirculation with young leachate. Valencia et al. (2009) investigated the addition of septic tank sludge to bioreactor landfills and observed that it increases the performance by increasing the moisture retention capability and supplies sufficient bacterial population for methanogenesis. It also acts as a sink for safe disposal of septic tank sludge. Yaman (2019) conducted a test by injecting water in four pilot scale bioreactor landfills containing different combinations of sewage sludge, MSW and geotextile layer. He observed that the reactor containing a mix of MSW and sewage sludge with one layer of geotextile provides more stabilization and less COD (Chemical oxygen demand) than other reactors.

Ayub (2019) studied the effect of addition of sludge and wastewater recirculation from Trinity river wastewater treatment plant. He observed that the injection of sludge

Table 1 Advantages of addition of biosolids in leachate recirculation

Liquid with biosolids/liquid alone for leachate recirculation	Advantage	References
Active sludge	In situ contaminant removal from leachate	Jun et al. 2007)
Sludge with saline water	Alternative to fresh water supplement	Alkaabi et al. 2009)
Septic tank sludge	Enhances the stabilization by increasing the moisture retention capacity of bioreactor	Valencia et al. 2009)
Leachate addition with municipal waste water treatment sludge	Accelerate biodegradation, results in faster stabilization	Rendra et al. 2007)
Wastewater from different industries such as fishery, brewery, and dairy	The methane generation observed in recirculation with waste water from fishery, brewery is relatively higher compared to dairy	Ko et al. 2012)
Addition of sludge with wastewater from Trinity River wastewater treatment plant	Increased level of methane generation with addition of biosolids compared to no recirculation	Ayub 2019)
Surface water, waste water effluents from industrial and commercial Addition of sewage sludge	To dispose industrial waste effluents effectively. To compensate for moisture in regions which face water scarcity Enhance the process of methanogenesis	Chembukavu et al. 2019)
Blended leachates	Enhance methane gas generation	Aromolaran and Sartaj 2021)

mixed with leachate has the highest methane production, followed by sludge with wastewater recirculation. In case of arid regions, wherein there is an absence of fresh water, saline water may also be used for moisture supplementation. Alkaabi et al. (2009) investigated the addition of saline water with sludge, and observed that it enhanced the rate of decomposition of MSW and about 14% higher yield of methane was generated in comparison to without addition of sludge.

The supplementation of inoculum due to addition of animal origin materials or sewage sludge reduces the time for stabilization because it enhances the degradation process by prolific bacillus activity (Callaghan et al. 2002; Sosnowski et al. 2003; Jun et al. 2007). Ko et al. (2012) suggested the utility of wastewater from various industries such as fishery, dairy and brewery act as an alternative source of influents to bioreactor landfill for leachate supplementation. Bioreactor landfill in the city of Calgary works towards sustainable development by utilizing liquid from adjacent storm water pond. The generated landfill gas is used as an energy resource by collecting it in a controlled manner without causing adverse impact on the environment (Hunte et al. 2011).

In recent advancement, attempts are made to upgrade the landfill gas collection system and utilize the landfill leachate for energy and fuels (Nanda and Berruti 2021). The utility of mature landfill leachate provides extra substrate and added nutrients to immature landfills. The supplementations of methanogens through injection of old leachate results in the improvement of overall production of methane in young landfills (Suna Erses and Onay 2003; Šan and Onay 2001; Yuen 2001).

2.2 *Landfill Gas*

A bioreactor landfill can yield landfill gas in a very short time in comparison to a dry landfill. Gas production in landfills results from thermo-chemo-biological reactions. It is necessary to install pipe systems for gas extraction during filling condition or in its operation life for efficient gas collection and to avoid odor problems. MSW landfills act as the third largest source of methane emission in the United States, which accounts to about 15%, as shown in Fig. 3. This is equivalent to greenhouse gas emissions obtained from more than 21.6 million passenger vehicles. Therefore, the collection of greenhouse gas and their utilization as an energy resource can be considered as a sustainable solution. The captured gas is further processed to produce heat or fuel and sold in the market as renewable power source, and generates revenue which turns it into an asset from waste (US, E.P.A. 2022a).

The gas formation mechanism in sanitary landfills takes place due to anaerobic degradation of organic material which involves the following mechanisms as shown in Fig. 4. Fresh waste deposits contain huge amount of organic matter due to which rapid anaerobic fermentation occurs. This results in the formation of volatile compounds and hydrolysis of waste leads to generation of amino acid, sugars, fatty acids, which further degrade to intermediate products. Acid fermentation is accelerated by raising moisture content in early stage and further methanogenic phase takes place once landfill becomes mature (Kumar et al. 2011; Demirel and Scherer 2008).

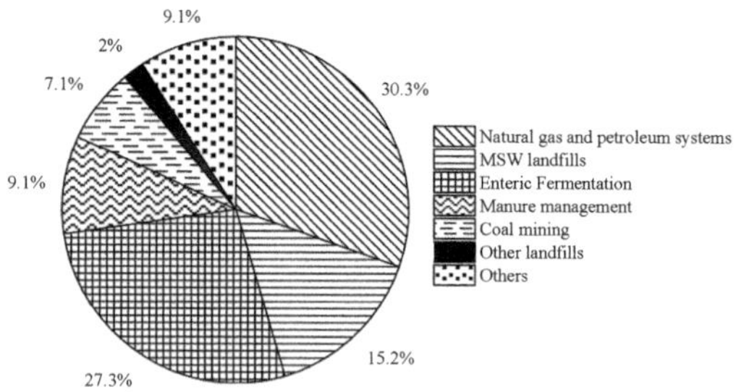

Fig. 3 Emissions of methane from various sources

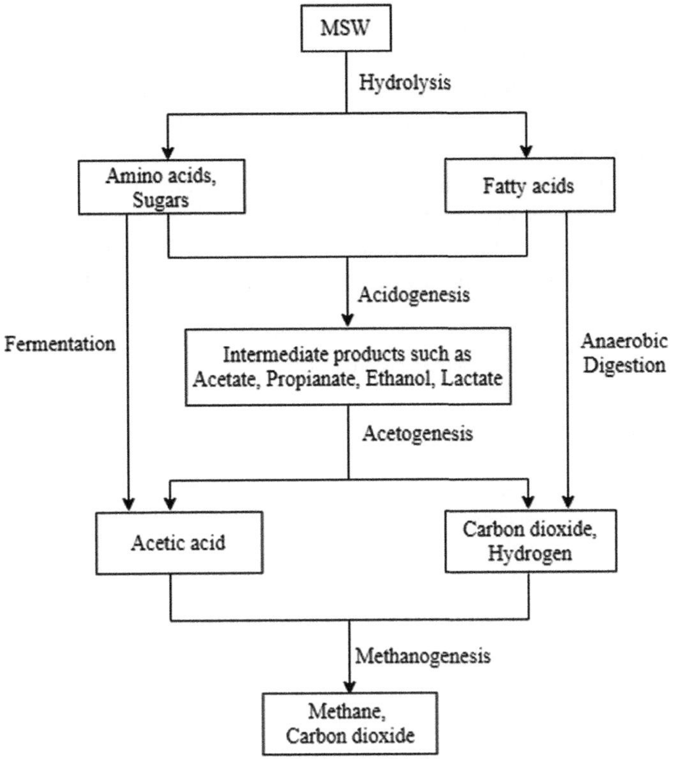

Fig. 4 Mechanism of gas formation in bioreactor landfills

Increased temperatures in landfills produces gases from volatile mixtures such as naphthalene and alcohol present in MSW. The presence of methanogenic bacteria is relatively more in landfill bed that makes the waste devoid of oxygen, and favors anaerobic condition (Nanda and Berruti 2021). Landfill gases comprise mainly of methane (40–60%) and carbon dioxide (30–50%) and about 10% of other pollutants (Mali Sandip et al. 2012; Eklund et al. 1998; Ehrig et al. 2011; Demirbas 2011). Since methane and carbon dioxide are main components present in landfill gas, they are major contributors to global warming. The potential of methane is 25 times more as compared to carbon dioxide for favoring global warming (Nanda et al. 2016).

Landfill gas composition depends upon properties and percent of biodegradable content in solid waste, rain infiltration, moisture content, oxygen content, pH, decomposition phase (Davoli et al. 2010). Other factors such as acidity, inhibitors and high percent of lignin have adverse effect on degradation (Jayasinghe 2013). The range of factors for favoring the biodegradation are shown in Table 2. However, the exposure of these gas emissions poses health hazards to human beings due to unpleasant odors and long-term exposure. Even frequent exposure of these gases leads to abnormalities (Narayana 2009). The consequences take place because of uncontrolled release of gas from landfills into atmosphere. So, to counter it, a gas collection is to be installed early in the methanogenic phase for early recovery (Amini et al. 2012). Table 3 shows the utilization of gas as energy resource from full scale landfills (US, E.P.A. 2012, 2022b). The currently operational energy generation projects convert landfill gas into various forms of energy such as electricity, upgrading landfill gas and medium-Btu gas.

The waste management generate revenue by producing power during methane combustion. This is a continuous process and involves numerous machines to generate electricity from biogas (Consonni and Viganò 2011; Benato et al. 2017). Incidentally, researchers have already carried studies to estimate the rate of gas

Table 2 Factors controlling the landfill gas production

Factor	Influence	Range	References
Density of waste	Increased waste densities decreases gas production due to acid accumulation; in turn reduces pH and inhibits methanogenesis	NA	Al-kaabi 2007)
pH	Low levels of pH inhibit the activity of methanogenic bacteria	6.8–8 6–8	Kumar et al. 2011; Warith 2003) Ehrig 1983)
Temperature (^{0}C)	The temperatures inside the landfill rises the activity of bacteria, ultimately results in increase in production of gas	30–40	Barlaz et al. 1990)
Water content (%)	Moisture content enhances the microbial decomposition results the production of landfill gas	$>=50$	Warith et al. 2001) Rees 1980)
Composition of waste (%)	Presence of more content of biodegradable fractions in landfill waste produces more amount of gases	30–50 40–60	Dasgupta and Mondal 2012) Sharma and Jain 2019)
Waste age (years)	Fresh wastes of recently buried produces more gases compared older waste	Age $<=$ 5–7	ATSDR 2022)
Nutrients	It is required to maintain nutrient level in the landfill for microorganism's activity	Organic matter to nitrogen to phosphorous 100:0.44:0.08	Chembukavu et al. 2019; Al-kaabi 2007)

Table 3 Landfill gas utilization from full-scale landfills

Landfill	Energy resource
Ämmässuo landfill, Espoo, Finland	Generates electricity of 15 MW utilized for nearby industries
Fort Benning Landfill, Georgia	Provides electricity for the onsite utility of U.S. army
Gaoantun Landfill, Beijing, China	Landfill gas used as a fuel for boilers to produce hot water or steam. The produced steam is used to generate electricity
Khelmintsky Landfill, Ukraine	Infrared heaters installed in landfill garage to convert landfill gas energy to heat
La Crosse County Landfill, Wisconsin	Landfill gas produces green power to local grid and for Gundersen's Onalaska campus to heat buildings and water
Loma Los Colorados landfill, Santiago, Chile	Generates electricity of 2 MW in 2009 and increased to 21.78 MW in phases
Millersville Landfill, Anne Arundel County, Maryland	Generates 3.2 MW of green power from landfill gas
Olinda Alpha Landfill, Orange County, California	Generates electricity of 32.5 MW
São João Landfill, São Paulo, Brazil	Generates electricity of 22 MW in 2007
Star Ridge Landfill, Alabama, USA	Landfill gas is utilized as a fuel for Jenkins Brick manufacturing unit
Three Rivers Regional Landfill, South Carolina, USA	Uses landfill gas as a fuel for the paper mill in beech island
Yancey-Mitchell County Landfill, North Carolina, USA	Landfill gas is used to generate renewable energy to run glass furnaces and pottery kilns Landfill gas is utilized to provide power supply to two craft studios, gallery, a visitor center, and heat to four greenhouse gases

generation in conventional landfills and suggested it is required to design and install gas collection system (Bogner and Matthews 2003; McBean 2011). The innovative features of bioreactor landfill eliminate the negative impacts on environment associated with disposing waste in landfills focusing on minimizing the emissions of greenhouse gases and maximize the recovery of energy (Hettiaratchi et al. 2021).

2.3 *Landfill Mining*

This concept was started in 1953 in Hiriya landfill in Israel to recover decomposed wastes (Zee et al. 2004). The MSW generated is stabilized inside the bioreactor landfill and is determined by studying the physical, chemical and biological characteristics including variations of spatial and temporal of leachate, landfill gas and solid waste (Warith et al. 2001; Morris et al. 2003). Landfill mining and its reclamation counteract socio-environmental problems such as groundwater contamination due to heavy metals and high toxic leachates (Krook et al. 2012). This concept was recognized as sustainable material management by merging waste management with material cycle (Krook 2010). Mined residue has been proclaimed as innovative strategy for waste deposits and it is necessary to have essential knowledge in feasibility of landfill mining (Hogland 2002). Wolfsberger et al. (2014) suggested that mining deposits created due to anthropogenic are employed as raw material fines because of presence of recyclable materials.

The strong reason for moving towards landfill mining is due to scarcity of land for origin of new landfills, urban development, environmental liabilities such as pollution as well benefits associated with reclaimed materials and revenue generation with mined residues (Zee et al. 2004). It not only reduces the footprint of landfill but also recovers the valuable materials such as steel, iron, aluminum, copper and tin (Johansson et al. 2017). Jones et al. (2013) estimated the range between 150,000–500,000 of active and closed landfills which are delivering significant amount of secondary materials in Europe.

Several activities included in landfill mining are excavation, sorting and segregation of waste into different fractions for different processes such as mechanical and biological treatment, utilization, disposal (Jones et al. 2013; López et al. 2018). The byproducts obtained from landfill mining are metals, glass, rubber, fabric, plastics, cardboard, soils, decomposed organic matter are termed as mining residues and their energy recovery are shown in Table 4. The recovered metals are recycled, polymeric substances are converted to energy since they have high calorific value, and decomposed organic materials are used as fertilizers for agriculture (Chembukavu et al. 2019; Nanda and Berruti 2021; Zhou et al. 2015a). The identification of heavy metals in landfills during mining is due to disposing of industrial wastes, mine wastes, hazardous household components such as paints, batteries, dyes and these are rich in metallic elements such as cadmium, iron, copper, nickel, lead, zinc (Jun et al. 2007).

The mined residues from bioreactor landfill contains rich nutrients, making it feasible for use as organic fertilizers in agriculture and as daily cover application in landfills (Chembukavu et al. 2019; Prabpai et al. 2009; Masi et al. 2014). The fine fraction from landfill mining can be mixed with soil to improve its nutrient content (Zhou et al. 2015b; Rong et al. 2017), and can alternatively acts as construction material (Quaghebeur et al. 2013). The residues of MSW requires composition elemental analysis to identify the availability of micro and macro nutrients for agriculture application (Soumaré et al. 2003). Özkaraova and Oral (2022) investigated the reuse of the residues as potential reactive barriers because of low costs. Jain et al. (2005) classified that three-fourth of the overall mined residue is soil and degraded matter based on the size of mining materials and the type of fractions. In Collier landfill, Florida, the recovered soil fractions from landfill mining are about 60% by weight.

Production of Refuse Derived Fuel. Refuse Derived Fuels (RDF) are obtained from several processing units by modifying the MSW composition with improved features of thermochemical processes (McHenry 2009). The plastics from landfill mining are used as sources for generating fuels (Zhou et al. 2014). Plastics reclaimed

Table 4 Material and energy recovery from landfill mining

Recovered material	Utility
Soil like fraction	Used as cover soils for landfills Organic fertilizer for agriculture
Plastics	Produce refuse derived fuels Plasma gasification Incinerate to generate heat or electricity
Metals	Recycled to regenerate metals Generally, aluminum and tin are highest recovery revenues Corroded ferrous metals are recycled Recycled metals use less energy for making products
Glass	Glass is recycled mainly used as cullet for making bottles, containers
Textiles	Recovered textile can be recycled into new raw materials
Stones and Construction waste	Regenerate to reuse as construction materials
Binders with high value	Energy recovery from solid recovery fuel treatment in gas plasma
Rubber	Recycling of rubber is cheaper and requires less energy
Other residuals require further disposal	Landfilled again

from MSW are incinerated to recover thermal energy to produce electricity. Also, these plastics serve as raw materials for derived fuel production (Al-Salem et al. 2009). It is required to reduce the emission of carbon from fossil fuels for mitigating the global environmental problems. Currently, the industries are not focusing on the utilization of RDF because it is an expensive fuel in comparison to oil and coal. RDF can acts as an alternative source of energy to reduce the carbon emissions (Chavando et al. 2022). Thus, RDF can serve as a feedstock for the generation of energy by replacing fossil fuels and can minimize the MSW in landfills.

Fine Fraction from Municipal Solid Waste. This study mainly focusses on the utilization of fine fractions of MSW finer than 4.75 mm size as a replacement of soil in geotechnical applications as backfill/embankment/subgrade material. This soil like material can be used for brick manufacturing/concrete as construction material. The basic geotechnical properties of fine fraction in MSW are reported in Table 5 (Rawat and Mohanty 2022). It is also noticed that about 60–70% of waste is categorized as fines (Havangi et al. 2017; Rawat and Mohanty 2021). Somani et al. (2018) studied the old dumpsites in India and reported that the mined residue contains about 50–60% of soil like material.

They investigated the feasibility for application of these mined residues in geotechnical purposes. The utility of reclaimed soil based on full-scale landfills is shown in Table 6. The development status of landfill mining is due to its potential advantages such as reclamation of space for reuse, recovered fractions of soils for utilization,

Table 5 Geotechnical characteristics of soil like fraction in MSW

Geotechnical characteristics	Values	References
Sand (%) Silt + clay (%) Specific gravity c (kPa) Φ (degree)	15–50 20–32 2.2–2.58 5–25 12–36	Datta et al. 2021)
Dry unit weight (kN/m^3)	8.8–16.5	Somani et al. 2018)
Water content (%)	7.4–40	Somani et al. 2018)
Liquid limit (%)	48–62	Oettle et al. 2010)
Plastic limit (%)	24–32	Oettle et al. 2010)

the recovered organic materials as refuse derived fuel and reduction of costs associated with landfill closure (Krook et al. 2012; Jones et al. 2013; Environmental Alternatives: Landfill mining: preserving resources through integrated sustainable management of waste 2022). Landfill mined soil fraction may be used as a man-made resource, which may be regarded as a sustainable solution to infrastructure development (Chandana et al. 2021).

3 Discussions

Sustainable landfill refers to the application of methods to reduce the pollution potential through in-situ remediation. In modern landfills, the pollutants are brought down to harmless substrates by flushing them out by organic matter stabilization and by accomplishing methane extraction (Jones et al. 2013). It is required to establish relationships with researchers from waste management and industry for the development of new technologies for sustainable operation of bioreactor landfills.

The climatic conditions can significantly affect the volume of leachate production, wherein warm weather conditions usually produce more leachate than colder ones (Mohd-Salleh et al. 2020). In cold weather conditions, solar and electric energy can be used for heating the leachate prior to circulation to control the ambient temperature inside the landfills (Zico et al. 2021). Continuous mode of liquid addition to landfill yields larger moisture distribution zones. However, intermittent addition is preferred in cases of liquid scarcity to minimize the quantity of liquid for recirculation. The operation of bioreactor landfills is associated directly with liquid injection process. However, the liquid addition is influenced by various other governing parameters such as pH neutralization, sludge addition, temperature control, nutrient addition, recirculation flow rate, waste shredding, and addition of mature leachate. There exists a research gap in the inter-dependence of these parameters, and their overall influence on waste stabilization. The addition of biosolids along with supplementing liquids for enriching the leachate are the new areas of development in bioreactor technology. Based on the characteristics of the produced leachate, it is recommended to have

Table 6 Use of recovered soil like fraction from landfill mining projects in full scale landfills

Landfill	Particle size (mm)	Soil like material (%)	Suggested reuse	References
Alachua County Southwest Landfill, Florida, USA	< 0.425	44.5	Daily cover	Jain et al. 2005)
Burghof landfill, Germany	NA	70.5	Reused at landfill site	Somani et al. 2018)
Burlington County landfill, New Jersey, USA	< 25.4	50	Daily cover	Hull et al. 2005)
Edinburg landfill, UK	25	75	Cover, beneficial use in off site	Dhar 2015)
Filbornia landfill, Sweden	18	55–70	Covering material	Somani et al. 2018)
Frey farm landfill, Canada	25	41	Cover material	Somani et al. 2018)
Kudjape landfill, Estonia	< 40	54	Daily cover	Bhatnagar et al. 2017)
Nanji-Do landfill, Seoul, Korea	NA	NA	Road construction material	Song et al. 2003)
Naples landfill, USA	75	40–60	Cover material	Somani et al. 2018)
Perdido landfill, USA	25	70.4	Daily cover and also reuse outside the landfill for various purposes	Dhar 2015)

integrated treatments before leachate disposal. Proper estimation of design of liquid storage and flow rates for supplementation into waste must be developed to ensure sufficient storage for leachate management.

The gases generated from bioreactor landfills primarily consist of methane, carbon dioxide, and minor proportions of organic volatile chemicals and hazardous pollutants. The landfill gas collection and its conversion to energy poses a great challenge due to the uncertainties associated with the governing factors as listed in Table 3. An increase in the rate of leachate recirculation during the initial stages reduces the temperature inside the landfill, which might affect the gas production to some extent. The carbon dioxide is to be separated from the landfill gases and the sequestration is also a major challenge. The success of energy generation projects depends on various factors such as quality and quantity of gas and the distance of the plant from the location intended for its direct use to generate electricity. The efficient conversion of landfill gas into electricity plays a key role in future energy generation scenario. This is because the energy production is obtained by exploiting the huge quantities of waste and unused residues. On this account, the waste management strategy should be to estimate gas generation potential and define how government policies can keep up its sustainability and environmental continence.

The byproducts obtained after the treatment of mined residues that have more carbon fraction with high calorific value can be used for energy recovery. The recyclable metals have a good market value and can serve as raw materials for further processing. Landfill mining is still lacking in developing countries due to lack of awareness on overall characteristics of mined residue because of its heterogeneity. Even though the researchers have realized the importance of reuse of mined soil fractions, the unavailability of proper guidelines restricts its usage in geotechnical and geo-environmental applications. It is also important to consider the aspect of landfill mining during the design and operation of bioreactor landfills so that the mined wastes may be easily accessible. In the present study, the process of landfill mining has been highlighted as a key contributor for the development of sustainable strategies for waste management.

4 Conclusions

Landfills act as long-term storage houses of disposed MSW that may pose adverse environmental problems such as global warming, groundwater contamination and health effects. It has been realized that bioreactor landfills are one of the most preferred options for disposing the MSW. New technologies are evolving in bioreactor landfills to facilitate the faster degradation and enhancement of landfill gas production process by addition of biosolids with moisture supplementation. Waste water effluents from fisheries, brewery and diary industries are also suitable for leachate recirculation. Efforts are required to move forward to achieve a circular economy to use waste as a secondary resource, for recovery, reuse, and recycling. The landfill gas, which contains about 40–60% of methane, has a high potential for generation of energy. Presently, 70% of the operating "landfill gas to energy" projects in the United States are working to generate electricity. The energy recovery from the waste prevents around 200 kg of carbon dioxide per ton of waste diverted from solid waste landfills. Also, it replaces the utility of fossil fuels approximately by 2.5 million tons. The mined residues from landfill mining activities are turned into economical and energy resources. The landfill mining reduces the volume of landfill for land reuse and is highly relevant to implement in new landfills because after a certain time, the stabilized waste is easily accessible for mining. Based on literature, it is observed that 40–75% of the recovered material from mined residues is soil-like fraction, that has properties similar to non-cohesive soils. Therefore, with proper treatments, these materials may be suitable in geotechnical applications such as backfill/ embankment/ subgrade materials. Based on the several advantages associated with bioreactor landfills, it is time to focus seriously on its acceptance as a prominent strategy for deriving long-term regulatory, monetary, environmental and social benefits. The waste management has to be taken as a challenge in a way that would attract the attention of research community to plan and implement safe disposal of MSW and in parallel, focus on generation of revenue from the MSW by converting waste into a valuable resource.

References

Al-kaabi S (2007) Effect of salinity on biodegradation of MSW in bioreactor landfills. Ph.D. thesis, Carleton University, Ottawa, Ont., Canada

Alkaabi S, Van Geel PJ, Warith MA (2009) Effect of saline water and sludge addition on biodegradation of municipal solid waste in bioreactor landfills. Waste Manag Res 27:59–69. https://doi.org/10.1177/0734242X07082107

Al-Salem SM, Lettieri P, Baeyens J (2009) Recycling and recovery routes of plastic solid waste (PSW): a review. Waste Manag 29:2625–2643. https://doi.org/10.1016/j.wasman.2009.06.004

Amini HR, Reinhart DR, Mackie KR (2012) Determination of first-order landfill gas modeling parameters and uncertainties. Waste Manag 32:305–316. https://doi.org/10.1016/j.wasman.2011.09.021

Aromolaran A, Sartaj M (2021) Enhancing biogas production from municipal solid waste through recirculation of blended leachate in simulated bioreactor landfills. Biomass Convers Biorefinery 19–26. https://doi.org/10.1007/s13399-020-01260-5

ATSDR—Landfill Gas Primer—Chapter 2: Landfill Gas Basics (2022). https://www.atsdr.cdc.gov/hac/landfill/html/ch2.html. Last Accessed 31 Jan 2022

Ayub A (2019) Effect of sewage sludge and wastewater recirculation on the degradation of municipal solid waste in bioreactors. https://rc.library.uta.edu/uta-ir/handle/10106/28820

Barlaz MA, Ham RK, Schaefer DM, Isaacson R (1990) Methane production from municipal refuse: a review of enhancement techniques and microbial dynamics. Crit Rev Environ Control 19:557–584. https://doi.org/10.1080/10643389009388384

Benato A, Macor A, Rossetti A (2017) Biogas engine emissions: standards and on-site measurements. Energy Proc 126:398–405. https://doi.org/10.1016/j.egypro.2017.08.278

Berge ND, Reinhart DR, Townsend TG (2005) The fate of nitrogen in bioreactor landfills. Crit Rev Environ Sci Technol 35:365–399. https://doi.org/10.1080/10643380590945003

Bhatnagar A, Kaczala F, Burlakovs J, Kriipsalu M, Hogland M, Hogland W (2017) Hunting for valuables from landfills and assessing their market opportunities a case study with Kudjape landfill in Estonia. Waste Manag Res 35:627–635. https://doi.org/10.1177/0734242X17697816

Bogner J, Matthews E (2003) Global methane emissions from landfills: new methodology and annual estimates 1980–1996. Global Biogeochem Cycles 17. https://doi.org/10.1029/2002gb001913

Callaghan FJ, Wase DAJ, Thayanithy K, Forster CF (2002) Continuous co-digestion of cattle slurry with fruit and vegetable wastes and chicken manure. Biomass Bioenerg 22:71–77. https://doi.org/10.1016/S0961-9534(01)00057-5

Chandana N, Goli VSNS, Mohammad A, Singh DN (2021) Characterization and utilization of landfill-mined-soil-like-fractions (LFMSF) for sustainable development: a critical appraisal. Waste and Biomass Valorization 12:641–662. https://doi.org/10.1007/s12649-020-01052-y

Chavando JAM, Silva VB, Tarelho LAC, Cardoso JS, Eusébio D (2022) Snapshot review of refuse-derived fuels. Util Policy 74. https://doi.org/10.1016/j.jup.2021.101316

Chembukavu AA, Mohammad A, Singh DN (2019) Bioreactor landfills in developing countries: a critical review. J Solid Waste Technol Manag 45:21–38. https://doi.org/10.5276/JSWTM.2019.21

Consonni S, Viganò F (2011) Material and energy recovery in integrated waste management systems: the potential for energy recovery. Waste Manag 31:2074–2084. https://doi.org/10.1016/j.wasman.2011.05.013

Dasgupta BV, Mondal MK (2012) Bio energy conversion of organic fraction of Varanasi's municipal solid waste. Energy Proc 14:1931–1938. https://doi.org/10.1016/j.egypro.2011.12.1190

Datta M, Gupta G, Somani M (2021) Geoenvironmental considerations for bulk reuse of MSW residues from old dumps and WTE plants in geotechnical applications. Indian Geotech J 51:63–83. https://doi.org/10.1007/s40098-020-00491-7

Davoli E, Fattore E, Paiano V, Colombo A, Palmiotto M, Rossi AN, Grande MI, Fanelli R (2010) Waste management health risk assessment: a case study of a solid waste landfill in South Italy. Waste Manag 30:1608–1613. https://doi.org/10.1016/j.wasman.2009.10.013

Demirbas A (2011) Waste management, waste resource facilities and waste conversion processes. Energy Convers Manag 52:1280–1287. https://doi.org/10.1016/j.enconman.2010.09.025

Demirel B, Scherer P (2008) The roles of acetotrophic and hydrogenotrophic methanogens during anaerobic conversion of biomass to methane: a review. Rev Environ Sci Biotechnol 7:173–190. https://doi.org/10.1007/s11157-008-9131-1

Dhar A (2015) Landfill mining—a comprehensive review. University of Texas at Arlington. https://doi.org/10.13140/RG.2.2.31092.55687

Ehrig HJ (1983) Quality and quantity of sanitary landfill leachate. Waste Manag Res 1:53–68. https://doi.org/10.1016/0734-242X(83)90024-1

Ehrig H-J, Schneider H-J, Gossow V (2011) Waste, 7. Deposition. In: Ullmann's Encyclopedia of Industrial Chemistry, Wiley-VCH Verlag GmbH & Co. KGaA, Weinheim, Germany. https://doi.org/10.1002/14356007.o28_o07

Eklund B, Anderson EP, Walker BL, Burrows DB (1998) Characterization of landfill gas composition at the fresh kills municipal solid-waste landfill. Environ Sci Technol 32:2233–2237. https://doi.org/10.1021/es980004s

The Environment protection act: an umbrella legislation for environmental related laws (1986) Last Accessed 20 Jan 2022

Environmental Alternatives: Landfill mining: preserving resources through integrated sustainable management of waste (2022). http://www.enviroalternatives.com/landfill.html. Last Accessed 13 Jan 2022

Guérin R, Munoz ML, Aran C, Laperrelle C, Hidra M, Drouart E, Grellier S (2004) Leachate recirculation: moisture content assessment by means of a geophysical technique. Waste Manag 24:785–794. https://doi.org/10.1016/j.wasman.2004.03.010

Havangi V, Sinha A, Parvathi G, Chandra S (2017) Municipal solid waste in road embankment construction-a case study. J Indian Road Congr 78–90

Hettiaratchi JPA, Jayasinghe PA, Yarandy TA, Attalage D, Jalilzadeh H, Pokhrel D, Bartholameuz E, Hunte C (2021) Innovative practices to maximize resource recovery and minimize greenhouse gas emissions from landfill waste cells: historical and recent developments. J Indian Inst Sci 101:537–556. https://doi.org/10.1007/s41745-021-00230-8

Hogland W (2002) Remediation of an old landsfill site. Environ Sci Pollut Res 9:49–54. https://doi.org/10.1007/BF02987426

Hull RM, Krogmann U, Strom PF (2005) Composition and characteristics of excavated materials from a New Jersey Landfill. J Environ Eng 131:478–490. https://doi.org/10.1061/(ASCE)0733-9372(2005)131:3(478)

Hunte CA, Hettiaratchi JPA, Meegoda JN, Hettiarachchi CH (2011) Performance of a waste cell in cold climate operated as an anaerobic landfill bioreactor. Geo-Frontiers Congress 1276–1286. https://doi.org/10.1061/41165(397)131

Jain P, Kim H, Townsend TG (2005) Heavy metal content in soil reclaimed from a municipal solid waste landfill. Waste Manag 25:25–35. https://doi.org/10.1016/j.wasman.2004.08.009

Jayasinghe P (2013) Enhancing gas production in landfill bioreactors by leachate augmentation. Ph.D. thesis, University of Calgary, Calgary. https://doi.org/10.11575/PRISM/25903

Johansson N, Krook J, Eklund M (2017) The institutional capacity for a resource transition—a critical review of Swedish governmental commissions on landfill mining. Environ Sci Policy 70:46–53. https://doi.org/10.1016/j.envsci.2017.01.005

Jones PT, Geysen D, Tielemans Y, Van Passel S, Pontikes Y, Blanpain B, Quaghebeur M, Hoekstra N (2013) Enhanced Landfill Mining in view of multiple resource recovery: a critical review. J Clean Prod 55:45–55. https://doi.org/10.1016/j.jclepro.2012.05.021

Jun D, Yongsheng Z, Henry RK, Mei H (2007) Impacts of aeration and active sludge addition on leachate recirculation bioreactor. J Hazard Mater 147:240–248. https://doi.org/10.1016/j.jhazmat.2007.01.001

Kaza S, Yao LC, Bhada-Tata P, Van Woerden F (2018) What a waste 2.0: a global snapshot of solid waste management to 2050. The World Bank, Washington DC, USA. https://doi.org/10.1596/978-1-4648-1329-0

Ko JH, Townsend TG, Kim H (2012) Evaluation of the potential methane yield of industrial wastewaters used in bioreactor landfills. J Mater Cycles Waste Manag 14:162–168. https://doi.org/10.1007/s10163-012-0053-1

Krook J (2010) Urban and landfill mining: emerging global perspectives and approaches. J Clean Prod 18:1772–1773. https://doi.org/10.1016/j.jclepro.2010.07.005

Krook J, Svensson N, Eklund M (2012) Landfill mining: a critical review of two decades of research. Waste Manag 32:513–520. https://doi.org/10.1016/j.wasman.2011.10.015

Kumar S, Chiemchaisri C, Mudhoo A (2011) Bioreactor landfill technology in municipal solid waste treatment: an overview. Crit Rev Biotechnol 31:77–97. https://doi.org/10.3109/07388551.2010.492206

López CG, Küppers B, Clausen A, Pretz T (2018) Landfill mining: a case study regarding sampling, processing and characterization of excavated waste from an Austrian landfill. Detritus 2:29–45. https://doi.org/10.31025/2611-4135/2018.13664

Mali Sandip T, Khare Kanchan C, Biradar Ashok H (2012) Enhancement of methane production and bio-stabilisation of municipal solid waste in anaerobic bioreactor landfill. Bioresour Technol 110:10–17. https://doi.org/10.1016/j.biortech.2011.12.027

Masi S, Caniani D, Grieco E, Lioi DS, Mancini IM (2014) Assessment of the possible reuse of MSW coming from landfill mining of old open dumpsites. Waste Manag 34:702–710. https://doi.org/10.1016/j.wasman.2013.12.013

McBean EA (2011) In-situ estimation of the methane generation rate for a wet and highly organic solid waste landfill. Int J Environ Waste Manag 8:123–132. https://doi.org/10.1504/IJEWM.2011.04097

McHenry MP (2010) Research institute for sustainable energy (RISE) 2009 report. Report for the Department of the Environment, Water, Heritage and the Arts, for the activity period 1 January 2009 – 30 April 2010, Murdoch University, Perth, Western Australia, pp 1–140

Mohd-Salleh S, Shaylinda M, Othman N, Azizan M, Yashni G, Afnizan W (2020) Sustainability analysis on landfilling and evaluation of characteristics in landfill leachate: a case study. In: IOP conference series materials science engineering. vol 736. https://doi.org/10.1088/1757-899X/736/7/072002

Morris JWF, Vasuki NC, Baker JA, Pendleton CH (2003) Findings from long-term monitoring studies at MSW landfill facilities with leachate recirculation. Waste Manag 23:653–666. https://doi.org/10.1016/S0956-053X(03)00098-9

Nanda S, Berruti F (2021) Municipal solid waste management and landfilling technologies: a review. Environ Chem Lett 19:1433–1456. https://doi.org/10.1007/s10311-020-01100-y

Nanda S, Reddy SN, Mitra SK, Kozinski JA (2016) The progressive routes for carbon capture and sequestration. Energy Sci Eng 4:99–122. https://doi.org/10.1002/ese3.117

Narayana T (2009) Municipal solid waste management in India: from waste disposal to recovery of resources? Waste Manag 29:1163–1166. https://doi.org/10.1016/j.wasman.2008.06.038

Oettle NK, Matasovic N, Kavazanjian E, Rad N (2010) Characterization and placement of municipal solid waste as engineered fill. In: Global waste management symposium San Antonio, Texas, pp 1–10

Özkaraova EB, Güven Oral E (2022) Reuse of industrial residues/wastes as a sustainable solution for landfill leachate contaminated groundwater. Soil Water Res 17:36–44. https://doi.org/10.17221/71/2021-SWR

Pohland FG (1980) Leachate recycle as landfill management option. J Environ Eng Div 106:1057–1069. https://doi.org/10.1061/JEEGAV.0001108

Prabpai S, Charerntanyarak L, Siri B, Moore MR, Noller BN (2009) Effects of residues from municipal solid waste landfill on corn yield and heavy metal content. Waste Manag 29:2316–2320. https://doi.org/10.1016/j.wasman.2009.02.009

Quaghebeur M, Laenen B, Geysen D, Nielsen P, Pontikes Y, Van Gerven T, Spooren J (2013) Characterization of landfilled materials: screening of the enhanced landfill mining potential. J Clean Prod 55:72–83. https://doi.org/10.1016/j.jclepro.2012.06.012

Rawat P, Mohanty S (2021) Experimental investigation on MSW fine mixed with fibers: fiber reinforced waste. J Hazardous, Toxic, Radioact Waste 25:04021009. https://doi.org/10.1061/(asce)hz.2153-5515.0000609

Rawat P, Mohanty S (2022) Parametric study on dynamic characterization of municipal solid waste fine fractions for geotechnical purpose. J Hazardous Toxic Radioact Waste 26. https://doi.org/10.1061/(asce)hz.2153-5515.0000659

Rees JF (1980) Optimisation of methane production and refuse decomposition in landfills by temperature control. J Chem Technol Biotechnol 30:458–465. https://doi.org/10.1002/JCTB.503300158

Reinhart DR, McCreanor PT, Townsend T (2002) The bioreactor landfill: Its status and future. Waste Manag Res 20:172–186. https://doi.org/10.1177/0734242X0202000209

Rendra S, Warith MA, Fernandes L (2007) Degradation of municipal solid waste in aerobic bioreactor landfills. Environ Technol 28:609–620. https://doi.org/10.1080/09593332808618822

Rong L, Zhang C, Jin D, Dai Z (2017) Assessment of the potential utilization of municipal solid waste from a closed irregular landfill. J Clean Prod 142:413–419. https://doi.org/10.1016/j.jclepro.2015.10.050

Šan I, Onay TT (2001) Impact of various leachate recirculation regimes on municipal solid waste degradation. J Hazard Mater 87:259–271. https://doi.org/10.1016/S0304-3894(01)00290-4

Sharma KD, Jain S (2019) Overview of municipal solid waste generation, composition, and management in India. J Environ Eng 145. https://doi.org/10.1061/(ASCE)EE.1943-7870.0001490

Simões GF, Catapreta CAA (2013) Monitoring and modeling of long-term settlements of an experimental landfill in Brazil. Waste Manag 33:420–430

Somani M, Datta M, Ramana GV, Sreekrishnan TR (2018) Investigations on fine fraction of aged municipal solid waste recovered through landfill mining: case study of three dumpsites from India. Waste Manag Res 36:744–755. https://doi.org/10.1177/0734242X18782393

Somani M, Datta M, Ramana GV, Sreekrishnan TR (2019) Leachate characteristics of aged soil-like material from MSW dumps: sustainability of landfill mining. J Hazardous Toxic Radioact Waste 23:04019014. https://doi.org/10.1061/(asce)hz.2153-5515.0000452

Song Y-S, Yun J-M, Hong W-P, Kim T-H (2003) Investigation of solid waste soil as road construction material. Environ Geol 44:203–209. https://doi.org/10.1007/s00254-002-0746-1

Sosnowski P, Wieczorek A, Ledakowicz S (2003) Anaerobic co-digestion of sewage sludge and organic fraction of municipal solid wastes. Adv Environ Res 7:609–616. https://doi.org/10.1016/S1093-0191(02)00049-7

Soumaré M, Tack FMG, Verloo MG (2003) Characterisation of Malian and Belgian solid waste composts with respect to fertility and suitability for land application. Waste Manag 23:517–522. https://doi.org/10.1016/S0956-053X(03)00067-9

Suna Erses A, Onay TT (2003) Accelerated landfill waste decomposition by external leachate recirculation from an old landfill cell. Water Sci Technol 47:215–222. https://doi.org/10.2166/wst.2003.0649

Tchobanoglous G, Theisen H, Vigil S (1993) Integrated solid waste management: engineering principles and management issues. McGraw-Hill Inc., New York, USA

The World Bank: Trends in Solid Waste Management (2022). https://datatopics.worldbank.org/what-a-waste/trends_in_solid_waste_management.html. Last Accessed 25 Jan 2022

Townsend TG (1998) In: Landfill bioreactor design and operation. 1st edn. Routledge, New York, USA. https://doi.org/10.1201/9780203749555
US, E.P.A. (2012) International best practices guide for LFGE projects
US, E.P.A. (2022a) Basic Information about Landfill Gas | US EPA. https://www.epa.gov/lmop/basic-information-about-landfill-gas. Last Accessed 15 Jan 2022
US, E.P.A (2000) Overview of Greenhouse Gases | EPA. https://www.epa.gov/ghgemissions/overview-greenhouse-gases. Last Accessed 18 Jan 2022
Valencia R, den Hamer D, Komboi J, Lubberding HJ, Gijzen HJ (2009) Alternative treatment for septic tank sludge: co-digestion with municipal solid waste in bioreactor landfill simulators. J Environ Manage 90:940–945. https://doi.org/10.1016/j.jenvman.2008.02.007
Van Der Zee DJ, Achterkamp MC, De Visser BJ (2004) Assessing the market opportunities of landfill mining. Waste Manag 24:795–804. https://doi.org/10.1016/j.wasman.2004.05.004
Warith M (2002) Bioreactor landfills: experimental and field results. Waste Manag 22:7–17. https://doi.org/10.1016/S0956-053X(01)00014-9
Warith M (2003) Solid waste management: new trends in landfill design. Emirates J Eng Res 8:61–70
Warith MA, Smolkin PA, Caldwell JG (2001) Effect of leachate recirculation on enhancement of biological degradation of solid waste: case study. Pract Period Hazardous, Toxic, Radioact Waste Manag 5:40–46. https://doi.org/10.1061/(ASCE)1090-025X(2001)5:1(40)
Wolfsberger T, Hollen D, Sarc R, Aldrian A, Budischowsky A, Zoscher A, Pomberger R, Lorber KE (2014) Landfill mining-case study : characterization and treatment of excavated waste from Austrian sanitary landfill sites and estimation of the resource potential. ISWA World Congr, Sao Paulo, Brazil, pp 8–10
Yaman C (2019) Improvement of leachate quality and waste stabilization in pilot-scale bioreactor landfills containing geotextile filters and sewage sludge. Desalin Water Treat 158:41–58. https://doi.org/10.5004/dwt.2019.24161
Yuen STS (2001) Bioreactor landfills: do they work? Geoenvironment. In: 2nd ANZ conference on environmental geotechnics, Newcastle, Australia, pp 28–30
Zhou C, Fang W, Xu W, Cao A, Wang R (2014) Characteristics and the recovery potential of plastic wastes obtained from landfill mining. J Clean Prod 80:80–86. https://doi.org/10.1016/j.jclepro.2014.05.083
Zhou C, Gong Z, Hu J, Cao A, Liang H (2015a) A cost-benefit analysis of landfill mining and material recycling in China. Waste Manag 35:191–198. https://doi.org/10.1016/j.wasman.2014.09.029
Zhou C, Xu W, Gong Z, Fang W, Cao A (2015b) Characteristics and fertilizer effects of soil-like materials from landfill mining. Clean—Soil, Air, Water 43:940–947. https://doi.org/10.1002/clen.201400510
Zico MM, Ricci BC, Reis BG, Magalhães NC, Amaral MCS (2021) Sustainable ammonia resource recovery from landfill leachate by solar-driven modified direct contact membrane distillation. Sep Purif Technol 264. https://doi.org/10.1016/j.seppur.2021.118356

Identification of Critical Urban Drainage Flooding Points Using MIKE URBAN + : A Case Study of Rohtak, Haryana, India

Divyanshi Pandey and K. K. Singh

1 Introduction

Urbanization has taken speedy growth in India. Urbanization has triggered abundant challenges, among which urban flooding is becoming one of the most important challenges. Factors responsible for urban flooding are hydrological factors, meteorological factors, and human factors. Increased water demand, unsustainable development, increased wastewater production, discharge into rivers and lakes, reduced infiltration rate, and increased use of groundwater have led to reduction in groundwater level, reduction in base flow, and ecology imbalances. These are also the major contributors of urbanization's hydrological effects (NIUA 2016). Urban development configuration has caused a major change in prevailing runoff pattern. Thus, it is very important in the present scenario to address the major issues in the planning of urban drainage networks to render the effects of over flooding. Because of the non-uniform rainfall pattern caused by global warming, there has been a significant change in runoff patterns, it is critical to develop a strategy to manage and operate the storm water drainage network that will ensure the smooth operation of the sewer system in the current environment. To protect the urban areas, storm water drainage discharge needs to be planned and implemented in a structured manner.

One can control human factors like scarce or inappropriate maintenance of storm water drains, encroachment on water bodies and drains, inappropriate planning, etc. Deprived natural drainage, blocking of the drainage system, extreme weather events, and variation in the river flood plain are the main causes of urban flooding. There is software that is presently used to design the storm water sewer network such as EPA SWMM. It is the most basic software that is used for the modeling of a stormwater drainage system. MIKE URBAN is the urban water modeling software that is now in

D. Pandey (✉) · K. K. Singh
Department of Civil Engineering, NIT Kurukshetra, Kurukshetra, India
e-mail: divyanshi_62100028@nitkkr.ac.in

K. K. Singh and C. S. Prasad Ojha (eds.), *Sustainable Management of Land, Water and Pollution of Built-up Area*, Society of Earth Scientists Series,
https://doi.org/10.1007/978-3-031-56176-4_16

use for the modeling of collection systems as well as for water distribution networks for the analysis, management, and operation of storm water drainage and wastewater systems in separate and combined systems. Parameters such as workflow, constancy, suppleness, openness are the most important criteria that one can prefer to solve by this software.

An urban drainage network model consists of basically three parts that are the precipitation input, the hydrological surface and runoff processes, and the hydrodynamics of the pipe network. The purpose of designing of stormwater drainage system is to have a properly efficient network that can alter the problems that are catered by frequent flooding due to insufficient loading of sewerage at the time of heavy rainfall. The objective of the study is in identifying the critical overflowing nodes using MIKE URBAN.

2 Study Area and Data Used

Study Area: The projected area here is ROHTAK city in Haryana, India which has a geographical urban area of 68.8 sq. km and a population of about 446,164 (as per census report 2011). Figure 2 shows study area. The latitude of Rohtak is 28.895515, and the longitude is 76.606613. GPS coordinates are 28° 53′ 43.8540" N and 76° 36′ 23.8068" E. The district area falls in the Yamuna subbasin of the Ganga basin and is mainly drained by the artificial drain No. 8 which flows from north to south. The city receives an average annual rainfall of 592 mm spread over 23 days. Physiographically the area is flat terrain. The approach is basically to model and simulate the area with MIKE URBAN + for an effective drainage network of the concerned city. With the skewness of rainfall patterns due to climate change, there is the problem of flooding in the city and so far, the city needs a planned and effective drainage network. Creation of nodes (manholes) linking them by pipe network giving reference to RL values (elevation) as the flow considered is gravity flow.

Data Used: The AutoCAD file of sewer storm network of projected area (Rohtak city) showing different drains and sewer treatment plants which is used to obtain the locations of manholes and outlets for sewer runoff. SRTM DEM of 90 m resolution was downloaded for the study area to get the elevation of different nodes that are created in MIKE URBAN + .

One Dimensional hydrodynamic and rainfall-runoff modelling has been done in MIKE URBAN + for this daily rainfall data of August month taken from the India-WRIS site. QGIS tool is used for extraction of elevation values of the ground level of the manholes.

Model impervious areas ranged from 30 to 70% based on house density in that region. Time Area method was used to determine the net effective rainfall-runoff using some standard parameters provided by MIKE URBAN + software such as Initial Losses, Reduction factors which occur due to many reasons such as Interception loss, etc. In this model, a constant 0.9 mm initial loss from all catchments along

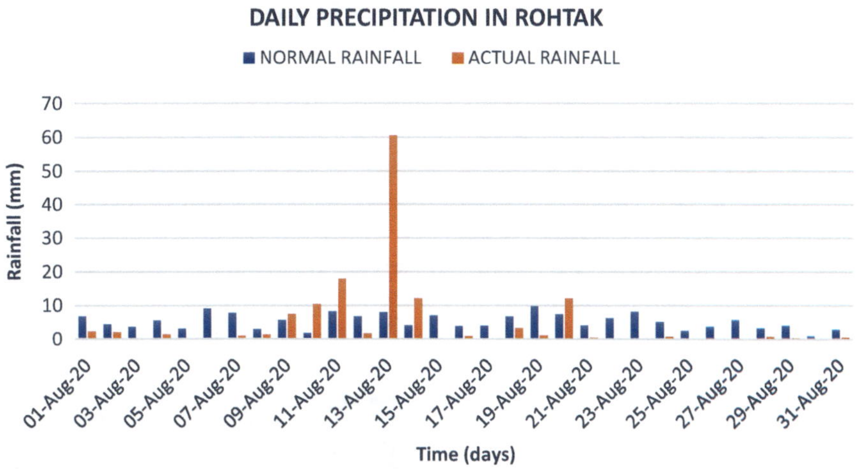

Fig. 1 Normal and actual precipitation in August 2020 in Rohtak. *Source* indiawris.gov.in/wris/rainfall

with a reduction factor of 0.85 is used, and the Time of Concentration was assigned as per the catchment size.

Rainfall Data: Extreme Rainfall events of August, 2020 has been used for study in Rohtak city. Figure 1 shows normal and actual precipitation.

3 Model Description

Hydrodynamic Module

The 1D engine solves the problem of water flowing through a network of nodes and structures. Boundary conditions at the network boundaries and initial conditions fully specify the problem (DHI 2017).

Saint–Venant Equations

It can be very difficult to deal with the problem of an arbitrary water flow through three-dimensional canals. But given that the geometry of pipes and canals permits specific sets of approximations, the problem space can be naturally fit and computational conclusions can be obtained. For purpose of approximation, Saint–Venant makes the following assumptions:

- Water is incompressible and homogeneous.
- The cosine of the angle with the horizontal is equal to 1 since the bottom slope of pipes and canals is modest.
- Wavelengths are substantial in relation to water depth which guarantees the water's parallel movement down the bottom.
- Water is not accelerated perpendicular to a reach's flow direction.

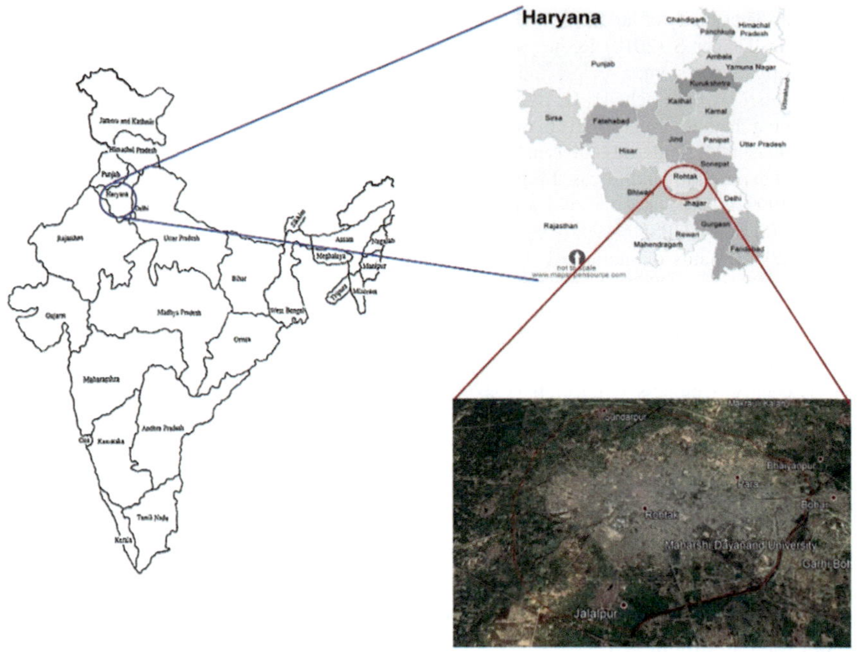

Fig. 2 Study area

Navier–Stokes' equations, often known as the Saint–Venant equations or Shallow Water equations, can be simplified to two equations using these presumptions.

$$\frac{\partial q}{\partial x} + \frac{\partial A_{fl}}{\partial t} = \mathrm{q}_{\mathrm{in}} \tag{1}$$

Equation (1) is called a mass equation or continuity equation and expresses the conservation of mass.

$$\frac{\partial q}{\partial t} + \frac{\partial\left(\alpha \frac{q^2}{A_{fl}}\right)}{\partial x} + gA_{fl}\frac{\partial h}{\partial x} + gA_{fl}I_f = \frac{\mathrm{f}}{\rho_{\mathrm{w}}} \tag{2}$$

Equation (2) is termed as a momentum equation that expresses the conservation of momentum.

Where

q = Discharge per unit length,

A_{fl} = Flow area, also called a cross-section area, q_{in} = lateral inflow (per length unit), h = water level, α = momentum distribution coefficient, I_f: = flow resistance term, in the form of friction slope, f = momentum forcing (per length unit), ρ_w = water density.

There are different formulations for flow resistance terms I_f which are Manning's, Darcy-weisbach, chezy, Colebrook-White for circular pipes, Hazen-Williams. Considering data availability Manning's is used to determine friction slope.

Manning's Formulation

$$I_f = fq|q| \tag{3}$$

With the friction factor:

$$f = \frac{1}{M^2 A_{fl}^2 R^{4/3}} \tag{4}$$

where M, manning resistance number. The equivalent value of M varies from 10 to 100.

Pressurized Flow

When pipelines or canals are full of water, the fundamental assumption used to derive the Saint–Venant equations is erroneous due to the transition from open surface flow to pressured flow. To include pressured flow in closed reaches, it might be reasonable to generalize the equations for free surface flow. The Preissmann slot, a hypothetical slot at the top of the reach, is used to accomplish this (DHI 2017).

Time Area Surface Rainfall-Runoff Model

In this model, the size of the imparting area, the initial loss and a continual hydrological loss all affect how much runoff occurs. The time-area (T-A) curve and concentration–time govern the runoff hydrograph's shape. Catchment reaction time and catchment form are determined by these variables.

Hydrological Parameters

- The quantity of precipitation required to cause surface runoff is shown by Initial Loss [m]. It entails flooding and stuffing catchment depressions, and it is a one-time loss.
- Hydrological Reduction is a runoff reduction factor accounting for water losses caused by such as evapotranspiration, and like imperfect imperviousness, etc. on the contributing area.
- When performing calculations, the available Time/Area curve is selected based on the shape of the catchment arrangement. The T/A curve is offered in three preconfigured forms: rectangular, divergent, and convergent catchment.
- A formula is introduced that covers curve shapes "between" the three common ones by using the time/area coefficient (DHI 2017).

$$(y = 1 - x)^{\frac{1}{a}} for 0 < a < 1 \tag{5}$$

$$(y = x)^{a} for 1 \leq a \tag{6}$$

where:

y = accumulated dimensionless area; x = dimensionless concentration time and a = time area curve coefficient.

4 Methodology

As the projected area is quite vast, so it is divided into smaller parts which provide better and accurate results. So, using the Catchment Delineation tool, generating catchments based on point layer (Thiessen Polygons) and opting area of interest to be complete Model network. Under the Catchment option, MIKE URBAN + software provides a catchment area along with its catchment number where one can find its Geometric Area and can provide population equivalent along with its impervious area in percentage corresponding to each catchment. Also, when rainfall-runoff data is taken into consideration, a hydrological model is needed in which simulation of model executed as per time area method.

The total number of Nodes (Diameter –1.5 m) used in this project were (2052) in which (22) nodes are an outlet (Diameter –2.0 m) and the remaining (2030) nodes are manholes. The nodes are connected by respective invert levels (bottom level). The difference between ground level and invert level was in the range of 0.8 –5.2 m. The diameter of the manholes (circular) is given as 1.5 m. The flow in the sewer network is mainly gravity flow so the nodes are connected from high elevation to lower elevation values while sometimes due to high runoff flow in network became pressurized flow which can be countered using proper assumptions.

In Fig. 3 process of surcharging of water level is shown. It also depicts ground as well as invert level of nodes along with pipeline flow (Schmitt et al. 2004). In this model, Links are used as Circular type with the following description of diameters for different link lines as given in Table 1.

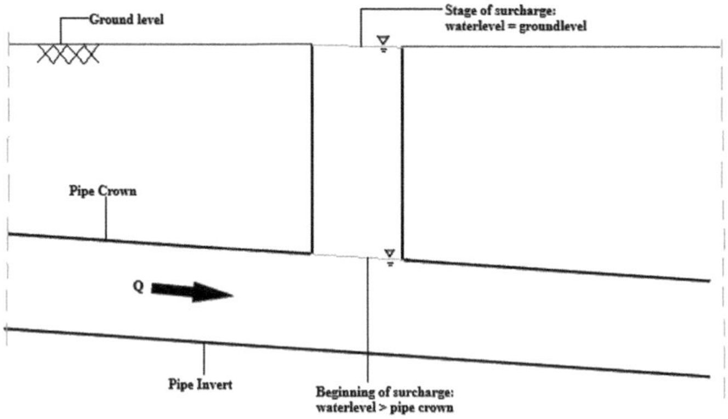

Fig. 3 Typical cross-section of pipe connecting with node

Table 1 Pipes dimension used in Rohtak sewerage system

Types of pipelines	Pipe diameter (m)
Trunkline	1.2
Mainline	0.9
Sewer line	0.6
Ordinary line	0.3

There are a total of 56 lines that includes 11 trunk lines, 22 main lines, and 23 sewer lines. After the connections of links next step were the delineation of catchments and their connections.

The drainage pipeline network's digitalized findings serve as the foundation for the sub-catchment division. The MIKE URBAN model's confluence node was positioned in the middle of the sub-catchment, and the study region's sub-catchments can be automatically divided using Thiessen polygons (Luan et al. 2018).

Processes of Model Construction is depicted in Fig. 4 as flow diagram which defines the basic steps to go through model construction:

MIKE URBAN + software provides a catchment area along with its catchment number where one can find its Geometric area and can provide population equivalent along with its impervious area in percentage corresponding to each catchment. Model impervious areas ranging from 30 to 70% based on house density in that particular

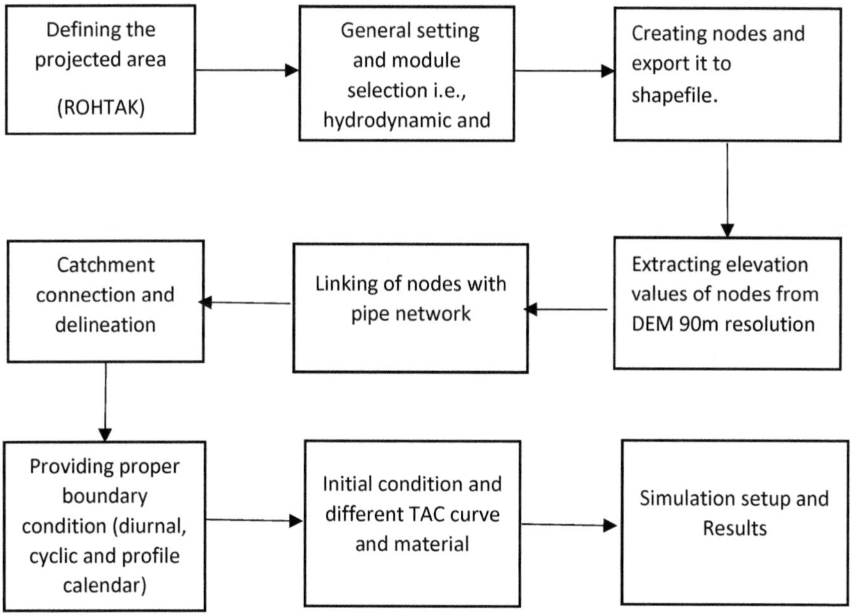

Fig. 4 Flow diagram of model construction

region (Alley 1983). Also, when rainfall-runoff data is provided then the Hydrological model method opted in which option for TIME-AREA method was preferred for this particular study area. The total number of catchments in this particular area were (2052). Catchment detailing provides information about every aspect of the catchment. It provides coordinates of catchment along with geometric area as well as catchment area from where one can calculate person equivalent. Also, provided a hydrological cycle to determine the curve for the time area method (DHI, "MIKE URBAN +" 2020).

Person Equivalent: Person Equivalent was calculated using the formula: PE = Round (Catchment Area*Model Impervious Area/100/0.014*5). Assuming 1500 square feet per house area (0.014 hectares) and 5 persons per house. Person equivalent calculated as per above formula found to be 1,170,029.

When Rainfall-Runoff data were considered the Time Area method was used to determine net effective rainfall-runoff using some standard parameters provided by MIKE URBAN + software such as Initial Losses, Reduction factors due to many reasons (Interception loss), etc. Infiltration losses are not considered under this model. In this study, constant 0.9 initial loss from all catchments along with reduction factor of 0.85 and Time of Concentration was set as a default parameter, and values were assigned by MIKE URBAN + software based on the catchment area.

Using a steady reduction of precipitation on impervious and semi-pervious surfaces, the runoff volume was determined after subtracting the initial loss. Calculating the temporal flow fluctuation on the surface involves measuring the amount of time that passes between when precipitation hits the surface and when it enters the main drainage pipe (Thorndahl and Schaarup-Jensen 2007). In this study, TA Curve1 was used to depict the Time Area curve.

5 Result and Discussion

With the help of Simulated res1d file result which is executable in MIKE URBAN software, analysis of Model is carried out based on Rainfall component as well as HD (Hydrodynamics) component. MIKE URBAN + provides results in the form of animation format for Catchment, Pipe connections, and Manholes.

Animation Analysis

- **Catchment Analysis:** The animation of Catchments shows the amount of rainfall-runoff in any catchment area at a particular time which plays an imperative role in the modelling of sewage flow system. Figure 5a shows total runoff in each catchment.
- **Pipes and Canals Analysis:** The animation of pipes and canals shows the amount of water flow in pipes which determines the type of flow in pipes i.e., gravity flow or pressurised flow. Sometimes the pressurised flow in pipes help in removing silt due to the high velocity of pressurized flow.

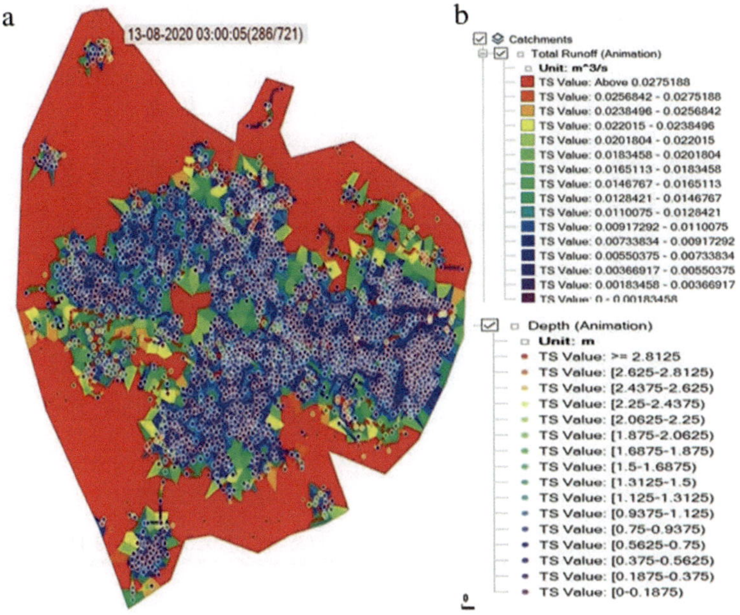

Fig. 5 **a** Total runoff in each catchment. **b** Depth of flow in each node

- **Nodes Analysis:** Node's animation shows the amount of water available in that particular manhole. With the help of an animation, overflowing Nodes can be determined which causes flooding in those areas. Figure 5b shows depth of flow in each Node.

Profile Plot Analysis

With the help of Profile plot, it is easy to analyse any part of the model using the Set Flag option. It provides a profile section of any chosen pipe flow path where one can check many options including Link Water Level and many more and with the help of animation of profile plot, one can identify overflowing nodes /Critical Nodes.

Profile Plot of trunkline and sewer line are presented in Fig. 6a and b respectively with analysis:

It is a profile plot of major TRUNK LINE which shows Link Water Level on the Maximum Rainfall. This profile shows that TRUNK LINE can hold that much rainfall. Hence, flooding would not occur due to manhole overflowing.

This is a profile plot of a SEWER LINE, where due to high rainfall some Nodes are overflown which led to Urban Flooding. So, with the help of the profile plot, these overflowing NODES are identified as Critical Nodes.

Overflowing Nodes: There are 58 nodes which show overflowing condition which led to urban flooding in ROHTAK city in the year 2020. These nodes are shown in Fig. 7 which also could be accessed by using the Google Earth link which is provided below:

a

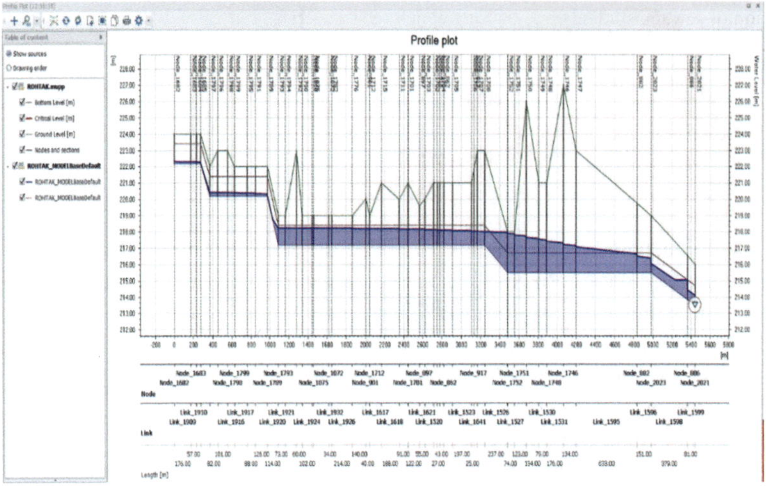

b

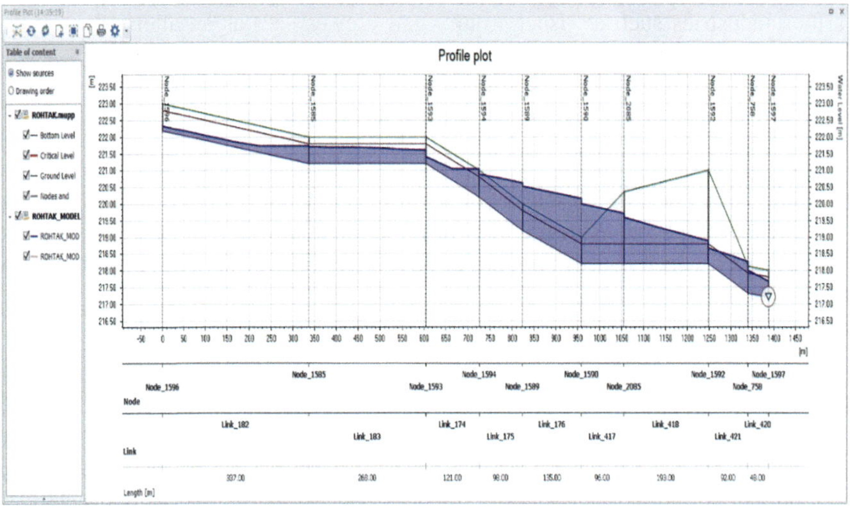

Fig. 6 **a** Profile plot of Trunkline. **b** Profile plot of Sewer line

https://earth.google.com/oneb/@28.90032561,76.58563194,223.28432964a,27622.01714369d,30y,-0h,0t,0r?utm_source=earth7&utm_campaign=vine&hl=e

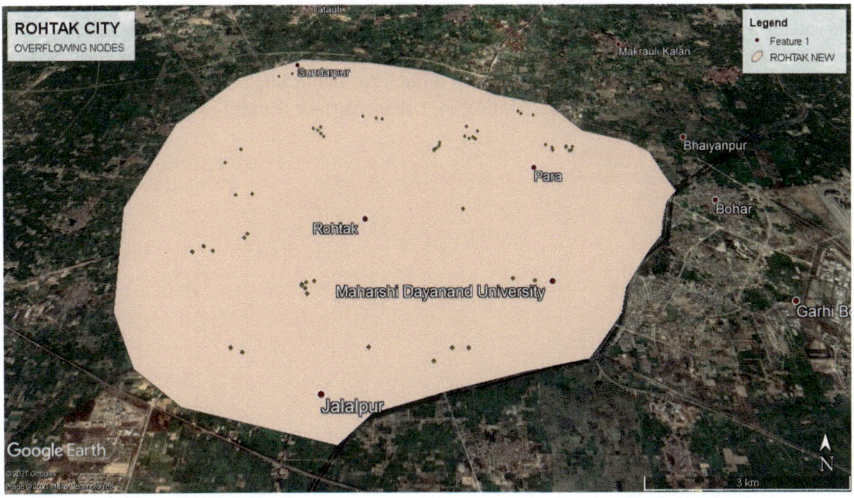

Fig. 7 Identification of overflowing nodes

6 Conclusions

In this Research paper, a 1-D hydrodynamic model was done to estimate urban flooding due to extreme rainfall and identify the critical points which led to the flooding. MIKE Urban + can quantitatively simulate complete High-resolution flood propagation that considers both sewage flow and runoff from rainfall. This model provides critical NODES which could be used to eliminate Urban flooding by using proper measures for example increasing pipe diameter or manhole size etc.

Acknowledgements Authors are extremely thankful to DHI (India) Water & Environment Pvt Ltd, New Delhi & the office of the Sub Divisional Officer, Public Health Engineering Department, Circle Rohtak, Haryana for providing MIKE Urban plus software and data, respectively.

References

DHI (2017) "Mike 1D Reference Manual." Simulation, pp 1–334

DHI, "MIKE URBAN +" (2020)

Alley JEVWM (1983) Effective impervious area in urban. Hydraul Eng 109(17669):313–319

Luan Q, Zhang K, Liu J, Wang D, Ma J (2018) The application of mike urban model in drainage and waterlogging in Lincheng county, China. Proc Int Assoc Hydrol Sci 379(2014):381–386. https://doi.org/10.5194/piahs-379-381-2018

NIUA (2015) Urban flooding, urban climate change fact sheet. 2016, [Online]. Available: http://www.parliament.uk/documents/post/postpn289.pdf. Accessed on 10 July 2015

Schmitt TG, Thomas M, Ettrich N (2004) Analysis and modeling of flooding in urban drainage systems. J Hydrol 299(3–4):300–311. https://doi.org/10.1016/j.jhydrol.2004.08.012

Thorndahl S, Schaarup-Jensen K (2007) Comparative analysis of uncertainties in urban surface runoff modelling. In: 6th international conference sustainable technology strategy urban water management, June 25–28, pp 1631–1638. [Online]. Available: http://vbn.aau.dk/files/44971436/Comparative_Analysis_of_Uncertainties_in_Urban_Surface_Runoff_Modelling_Abstract.pdf

Environmental Impacts of Gasoline Scooter and Electric Scooter

Yash Aryan and Anil Kumar Dikshit

1 Introduction

The motor vehicle-based transportation tremendously increasing in India for daily commute. The fast expansion in the number of automobiles has had a substantial negative impact on the human health and surroundings, since they are the primary contributor (20%) to urban air pollution. Furthermore, India's transportation sector accounts for 90% of total CO_2 emissions in metropolitan areas (Khandar and Kosankar 2014). Interestingly, over last 17 years from 2003 to 2019, the increase in number of scooters has been 79%. According to estimates, there were 296 million vehicles worldwide by the year 2019, of which 221 million (or 75%) were scooters (MoRTH: 2021). In urban areas, vehicle exhaust was responsible for 72% of NO_x and volatile organic compound (VOC) emissions and 87% of CO_2 emissions. Other gaseous pollutants and particulate matter (PM) were also released in significant amount from the vehicle exhaust (Guttikunda and Kopakka 2014). Globally, 35 Indian cities are among the top polluted cities based on PM 2.5 pollution. There is increase in global air pollution and electric vehicles are being introduced all around the world to reduce emissions from motor vehicles (Kumar and Padmanaban 2019).

The Indian government has also launched a number of programs to support and promote e-vehicles in order to conserve fossil fuels and reduce air pollution. Major businesses such as TATA and Mahindra have debuted their electric four-wheelers and Bajaj, Hero, Ather, and OLA have debuted their electric scooters in the markets of India. In India's major cities, the use of electric vehicles (mainly cars and scooters)

Y. Aryan (✉) · A. K. Dikshit
Environmental Science and Engineering Department, Indian Institute of Technology Bombay, Mumbai 40076, India
e-mail: yasharyan@iitb.ac.in

A. K. Dikshit
e-mail: dikshit@iitb.ac.in

K. K. Singh and C. S. Prasad Ojha (eds.), *Sustainable Management of Land, Water and Pollution of Built-up Area*, Society of Earth Scientists Series,
https://doi.org/10.1007/978-3-031-56176-4_17

is increasing, but obstacles include lack of infrastructure for battery charging and a shorter range per charge. With effect from April 2020, the Indian government mandated the manufacture and sale of BS-VI certified automobiles. The gasoline vehicles based on BS VI engines are expected to certainly reduce the vehicular emissions and hence reduction in urban air pollution. At the same time, electric vehicles are also gaining popularity. Therefore, there is necessity to investigate the ecological impacts of using gasoline and electric automobiles in India.

The objective of this study is to investigate the environmental implications of gasoline (petrol) and electric scooter use phases using a life cycle assessment (LCA) technique. LCA is globally used to evaluate a product's environmental impact across its entire life cycle. A few research on e-vehicles in India have been published in the literature (Monika 2019; Shinde et al. 2019; Hollingsworth et al. 2019; Ashnani et al. 2015).

2 Methodology

2.1 *Goal and Scope*

The current study attempts to assess the environmental consequences of using typical electric and gasoline scooters present in the Indian market. The analysis is restricted to the usage phase of both vehicles, which includes the production and consumption of gasoline (petrol) and the generation of electricity (in conventional coal thermal plants) in India. The study adopted BS-VI norms for gasoline scooter for proper assessment of environmental impacts. The impacts were calculated for 1000 km vehicle driven in average Indian conditions. The present study system boundary is shown in Fig. 1a, b. Fig. 1a inputs and outputs processes of the gasoline scooter and Fig. 1b presents the inputs and outputs processes of the electric scooter.

2.2 *Life Cycle Inventory*

A vital part of an LCA research is life cycle inventory (LCI), which entails obtaining and quantifying data on process input and output (Islam et al. 2016). The current study evaluated data on power requirements from electric scooter manufacturers, as well as data on gasoline mileage from scooter manufacturers. The tail pipe emissions from gasoline vehicles are based on the recently implemented BS VI standards. The data on gasoline and power generation adapted from the GaBi Professional database. The present study considered gasoline and electricity production data from Indian mix particularly. For a more accurate assessment of the environmental repercussions, the study took India-specific data into account. The electricity required to full charge electric scooter is in the range of 2–5 kwh as per different manufacturer. In this study,

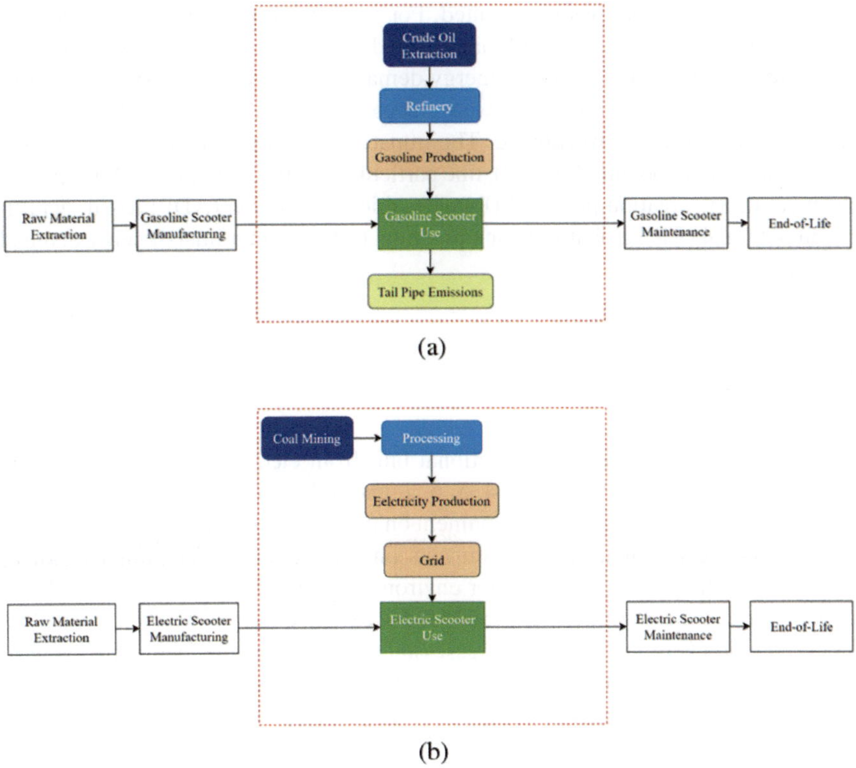

Fig. 1 **a, b** System boundary for the gasoline scooter and electric scooter respectively

electricity consumption of 4 kwh to full charge the electric scooter had been assumed and electricity is produced in coal power plant which is currently prevailing in India. The distance can be travelled is around 100 km per full charge of electric scooter as per the manufacturer information. According to Indian road and traffic circumstances, the majority of 125–150 cc scooters provide a fuel efficiency of 50 km per liter. Therefore, fuel efficiency was taken as 50 km per liter. For a trip of 1000 km, a total of 40 kwh of power and 20 L of gasoline were assumed to be consumed.

2.3 *Life Cycle Impact Assessment*

The impact assessment phase evaluates possible environmental impacts based on the data collected during inventory phase. In this phase of LCA study, the environmental burdens are converted into the possible impacts on the ecosystem. GaBi 10.5 software is used for assessing the impacts on the environment. In this study, a total of seven impact indicators were accounted. For impact assessment, one impact indicator taken from ReCiPe 2016 midpoint method, five impact indicators were adopted from CML 2001 method, and one energy demand category was considered. In this study, maximum impact indicators were selected from CML 2001 method because this method is accepted worldwide. The impact indicators accounted in this study are human toxicity potentials (HTP), fine particulate matter formation (PMF), photochemical ozone creation potential (POCP), abiotic depletion (ADP) fossil, acidification potential (AP), global warming potential (GWP) and primary energy demand (PED).

2.4 *Results and Discussion*

The environmental impacts per functional unit from electric scooter and gasoline scooter are presented in Figs. 2, 3, 4, 5, 6, 7, and 8. It is observed that the gasoline scooter had greater impacts on environment on three impact indicators while electric scooter had greater environmental impacts on the remaining four impact indicators. The gasoline scooter had greater environmental impacts on PED (817 MJ) in comparison to the electric scooter (556 MJ) due to high energy consumed in gasoline fuel production (Fig. 2). The impact assessment considered the whole life cycle of gasoline production (extraction, processing and transportation) at refinery in India. The ADP fossil impact category was more adversely affected by gasoline scooter (742 MJ) compared to electric scooter (435 MJ) due to use of gasoline as it is a fossil fuel (Fig. 3). The use of gasoline causes depletion in the fossil fuel resulting in higher impacts on abiotic depletion fossil impact category.

The electric scooter had a greater impact on the AP impact category (0.496 kg SO_2 eq.) than the gasoline scooter (0.112 kg SO_2 eq.), as shown in Fig. 4, since the BS VI rule has lowered the Sulphur content in fuel to 10 ppm and also during production of electricity sulfur dioxide gets emitted in coal power plant. The electric scooter had higher impacts on GWP impact category (42.5 kg CO_2 eq.) compared

Fig. 2 Environmental impacts on PED impact category

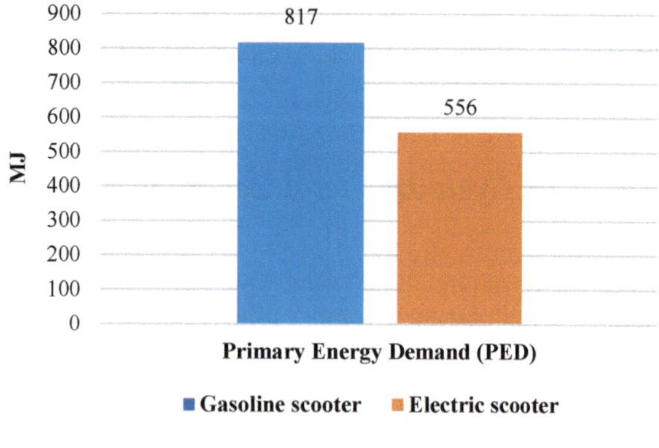

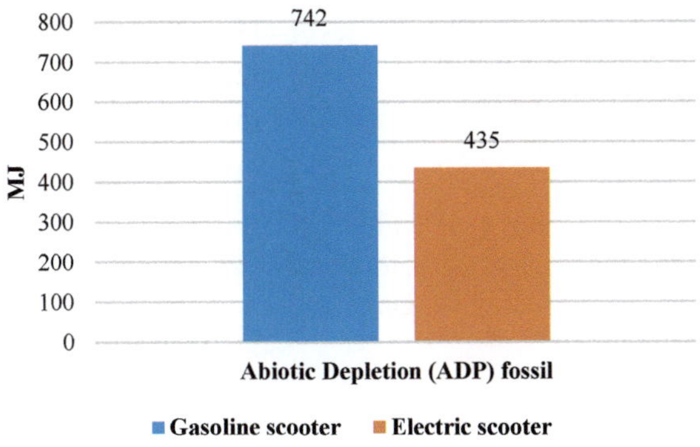

Fig. 3 Environmental impacts on ADP fossil impact category

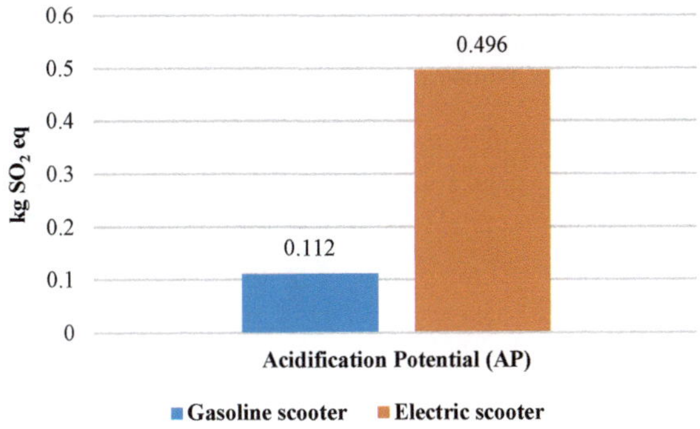

Fig. 4 Environmental impacts on AP impact category

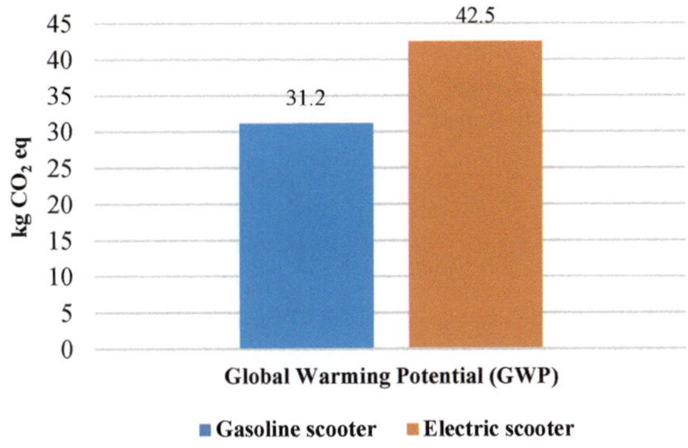

Fig. 5 Environmental impacts on GWP impact category

to gasoline scooter having 31.2 kg CO_2 eq. on GWP (Fig. 5). The greater impact on GWP by electric vehicle can be attributed to the fact that majority of electricity in India is produced from coal.

In thermal power plants in India. Existing LCA research comparing internal combustion engine vehicles (ICEV) with electric vehicles (EV) in developed nations like Sweden and Poland found that EVs had had less impacts on global warming

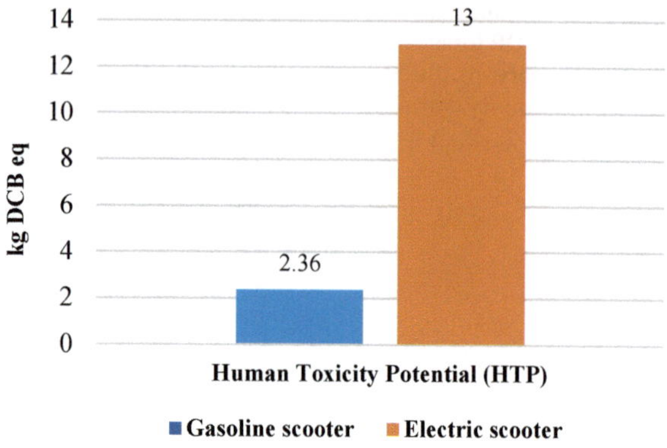

Fig. 6 Environmental impacts on HTP impact category

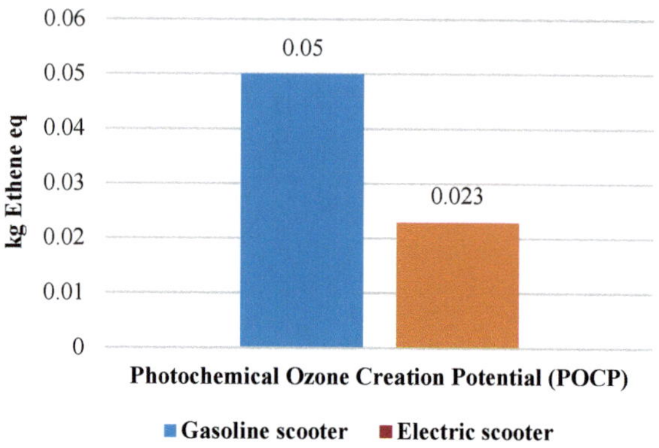

Fig. 7 Environmental impacts on POCP impact category

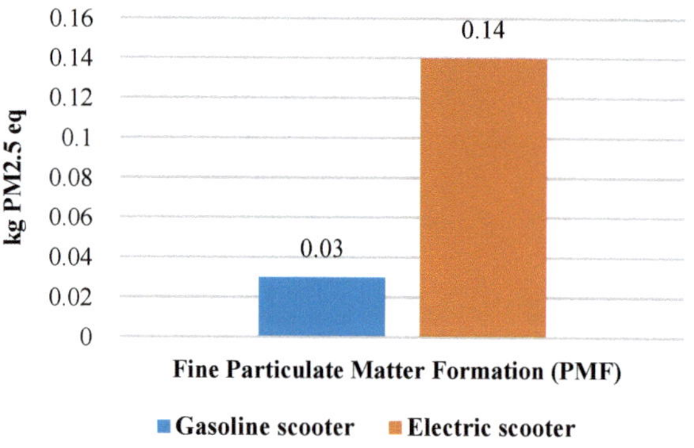

Fig. 8 Environmental impacts on PMF impact category

than ICEVs (gasoline or diesel) (Kurien et al. 2020). However, the carbon emissions from EV were found to be higher compared to GV for the current energy mix in India (Bare 2010). The carbon intensity of power generation in India is significantly higher (926 g CO_2/kWh) than the global average (542 g CO_2/kWh), resulting in larger

GWP impacts from EVs in India. (Shearer et al. 2017). However, India's crude oil upstream carbon intensity is lower than the average worldwide crude oil upstream carbon intensity, which is 10.3 g CO_2 eq/MJ (Hoekstra 2021). As seen in Fig. 5, electric scooter had higher global warming impacts which is attributed from the coal mining and generation of electricity from coal in thermal power plants. However, despite of high GWP impacts, electric vehicles should be encouraged and promoted as in its Nationally Determined Contribution (NDC), India has promised to develop enough non-fossil fuel-based power generation capacity to meet the nation's 50% demand by 2030. As a result, EVs will be more ecologically friendly than GVs in the future by 2030, and their effects on GWP will undoubtedly decline. EV(s) do not emit any tail-pipe emissions resulting in achieving positive public health benefits in urban areas as inhalation of pollutants emitted from gasoline vehicles pollutants cause's severe health effects (Grahame and Schlesinger 2010). The electric scooter had higher impacts on HTP impact category (13 kg DCB eq.) as coal is being for generating electricity in thermal power plants in India (Fig. 6).

The high impacts on human toxicity from electric scooter were attributed to mining and extraction of coal for generation of electricity. Coal mining and processing generates significant pollution by releasing toxic substances and various pollutants into the environment (Chabukdhara and Singh 2016). While, gasoline scooter affected more to POCP impact category (0.05 kg Ethene eq.) as shown in Fig. 7. As shown in Fig. 8, electric scooter had higher impacts on PMF impact category as in BS VI norms the particulate matter emissions has been limited to a very stringent limit.

The contribution of both the scooters has been shown in Fig. 9. The electric scooter contributed higher impacts compared to gasoline scooter on four impact indicators (AP, GWP, HTP and PMF) while gasoline scooter contributed more on remaining three impact indicators (PED, ADP fossil and POCP).

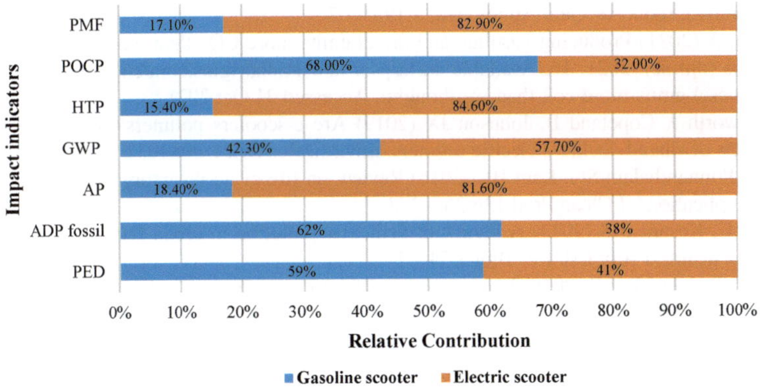

Fig. 9 Relative contribution of both the scooters on each impact category

2.5 Conclusions

The present study observed that the electric scooter had higher impacts on more impact indicators compared to gasoline scooter. The findings of this study are relatively comparable to those of Yu et al. (2018) as significant share of electricity at grid is obtained from coal in India and China. The gasoline (petrol) scooter had overall lesser impacts on the environment since the new BS VI established rules are extremely strict and BS VI engines are ecologically friendly generating very little pollutants and also the Sulphur level in fuel has been lowered to 10 ppm. The findings of this study conclude that in order to make electric vehicles more sustainable in the future, power for charging them must be produced from renewable sources like solar or wind. Additionally, coal and other non-renewable resources will be preserved. According to the study, the introduction of electric vehicles (EVs) for large vehicles like buses, Lorries, and three-wheelers like auto rickshaws is necessary in the near future since they produce the most pollution.

References

Ashnani MHM, Miremadi T, Johari A, Danekar A (2015) Environmental impact of alternative fuels and vehicle technologies: a life cycle assessment perspective. Procedia Environ Sci 30:205–210

Bare JC (2010) Life cycle impact assessment research developments and needs. Clean Technol Environ Policy 12:341–351

Chabukdhara M, Singh OP (2016) Coal mining in northeast India: an overview of environmental issues and treatment approaches. Int J Coal Sci TEchnol 3(2):87–96

Grahame TJ, Schlesinger RB (2010) Cardiovascular health and particulate vehicular missions: a critical evaluation of the evidence. Air Qual Atmos Health 3(1):3–27

Guttikunda SK, Kopakka RV (2014) Source emissions and health impacts of urban air pollution in Hyderabad. India Air Qual Atmos Heal 7:195–207

Hoekstra, A (2021) Producing gasoline and diesel emits more CO_2 than we thought. Innovation origins (online). 2021, 16. Available at: https://innovationorigins.com/en/producing-gasoline-and-diesel-emits-more-co2-than-we-thought/, Accessed 21 Oct 2021

Hollingsworth J, Copeland B, Johnson JX (2019) Are e-scooters polluters? The environmental impacts of shared dockless electric scooters. Environ Res Lett 14:084031

Islam S, Ponnambalam SG, Lam HL (2016) Review on life cycle inventory: methods, examples and applications. J Clean Prod 136:266–278

Khandar C, Kosankar S (2014) A review of vehicular pollution in urban India and its effects on human health. J Adv Lab Res Biol 5:54–61

Kumar R, Padmanaban S (2019) Electric vehicles for India: overview and challenges. IEEE India Newsl 14:139–142

Kurien C, Srivastava AK, Molere E (2020) Indirect carbon emissions and energy consumption model for electric vehicles: Indian scenario. Integr Environ Assess Manag 16(6):998–1007

Monika M (2019) A study on customer perception towards e-vehicles in Bangalore. J Emer Tech Innov Res 6

MoRTH (2021) MoRTH Annual report. https://morth.nic.in/sites/default/files/Annual%20Report%20-%202021%20%28English%29_compressed.pdf. Accessed 16 Sep 2021

Shearer C, Fofrich R, Davis SJ (2017) Future CO_2 emissions and electricity generation from proposed coal-fired power plants in India. Earth's Future 5(4):408–416

Shinde AM, Dikshit AK, Singh RK (2019) Comparison of life cycle environmental performance of public road transport modes in metropolitan regions. Clean Technol Environ Policy 21:605–624

Yu A, Wei Y, Chen W, Peng N, Peng L (2018) Life cycle environmental impacts and carbon emissions: a case study of electric and gasoline vehicles in China. Transp Res Part D Transp Environ 65:409–420

Review of Decentralized Wastewater Treatment Technologies and Systems

Harshvardhan Soni and Anil Kumar Dikshit

1 Introduction

Rapid urbanization has resulted in a scarcity of freshwater, with billions of people suffering from the unavailability of a proper sanitation system and lack of safe freshwater. This situation is worst in small rural areas. Around 40% of the population lacks a sanitation system and 25% of people lack access to those systems whereas about 82% of the rural population face this problem on a day-to-day basis (Chong et al. 2013; Ho 2005; Massoud et al. 2009). The adverse impacts of all these situations are well documented (Chong et al. 2013; Abegglena et al. 2008). At the same time, limited financial resources are the major impedance in the provision of environmentally and economically safe treatment systems (Fach and Fuchs 2010). Most of the time, centralized systems are unable to handle the large volume of sewage generated due to issues in collecting sewage from far away communities. Hence, there may be a need to look for decentralized or local systems as decentralized systems may be installed anywhere without having a huge budget and also decrease the fresh water demand by non-potable reuses of reclaimed water. The objective of the present paper is to review the various decentralized wastewater systems and technologies available as of now.

H. Soni (✉) · A. K. Dikshit
Environmental Science and Engineering (ESE) Department, Indian Institute of Technology Bombay, Powai, Mumbai 400076, India
e-mail: 204180005@iitb.ac.in

A. K. Dikshit
e-mail: dikshit@iitb.ac.in

K. K. Singh and C. S. Prasad Ojha (eds.), *Sustainable Management of Land, Water and Pollution of Built-up Area*, Society of Earth Scientists Series,
https://doi.org/10.1007/978-3-031-56176-4_18

2 Status of Wastewater Generation in India and Its Treatment

In India, 16,652 MLD of sewage is being generated from 299 class-I cities, out of this 59% is generated from the metro cities. Among these 299 class-I cities, only 160 cities have sewerage systems. The sewerage system used is either open, closed, or piping system, and 4037 MLD (24%) is being treated while the remaining (12,625 MLD) is discharged directly without any treatment (Dhote et al. 2012).

Need for Wastewater Treatment.

The adverse impacts of discharging untreated wastewater into the environment are as follows:

- As the untreated sewage is discharged into any water body, the solids present in the sewage produces foul gases after decomposition in anaerobic conditions from the bed of the water body.
- Untreated sewage will consume the DO already present in the river and result in septic conditions, which deteriorates aquatic life.
- Discharge of untreated sewage causes algal bloom in the river and causes eutrophication of the rivers.
- Untreated sewage contains pathogens and toxic compounds, which cause diseases (Dhote et al. 2012).

So, to overcome the above-mentioned reasons, the sewage is required to be treated before it is discharged into any water body.

2.1 Limitations of Centralized Wastewater Treatment Systems

Nowadays centralized treatment systems are sometimes unable to handle the large quantities of wastewater being generated from the cities. The sewer networks are also inadequate due to many reasons. Sometimes Geographic constraints also lead to the problem of laying in sewers varying or depending from place to place. Centralized systems are unable to produce sustainable outcomes in terms of social, economic, and environmental aspects. These systems consume much high energy and have a high risk of water-borne diseases. The discharge of untreated or partially treated waste from these systems can result in eutrophication in the water body in which it is discharged (Dhote et al. 2012).

3 Planning and Implementing Wastewater Reuse

3.1 Appropriate Technology

The following factors should be taken into consideration while selecting a technology for a decentralized system:

- Low cost, low energy, low mechanization to reduce the malfunctioning of the system (Boller 1997; Frijns and Jansen 1996).
- Simple in operation, local labor-intensive, not only relying on expensive chemicals, capable of recycling the resources (Boller 1997; Frijns and Jansen 1996; Mara and Cairncross 1989).
- Easily upgradable when required (Boller 1997).
- Public acceptance is the major key factor in any wastewater reuse project, and the rejection or non-acceptance by the public can jeopardize the treated sewage reuse (Asano and Levine 1996). The choice of any technology must be made well in advance so as the self-sufficiency and the capacity of the community can be analyzed, which means the treatment options can be handeled well by the users of the facility. (Boller 1997) Suggested that the technology selected should be such that it requires a low level of maintenance and control. The technology selected should be such that it should be capable of achieving pathogen removal for reuse purposes (Yu et al. 1997).

3.2 Reuse of Treated Wastewater

The applications of reclaimation of treated sewage includes various irrigation activities like highway plantings, public gardens, play areas, golf courses, and for flushing of toilets etc. (Chawathe and Kantawala 1987). Typical industrial applications include use in cooling systems, the food processing industries, and other high rate water uses (Bouwer 1993a, 1993b). Currently, in India, treated sewage is used in irrigation, toilet flushing, cooling of air conditioning systems, boiler feed water, and industrial processing water (Chawathe and Kantawala 1987). National Policy has been framed in china, which motivates the use of treated sewage in agriculture and other industrial and municipal uses (Zhongxiang and Yi 1991). In Japan, reclaimed sewage is used for flushing of water closets and urinals, industries, stream recharging, and flow enhancement. Classification of Decentralized Wastewater Treatment Technologies.

3.3 Natural Treatment Systems

Natural treatment systems are those treatment systems that make use of natural materials like soil, plants, microbes, air, sunlight for treatment. They are very cheap to construct and also have low operation and maintenance costs, they can be used where a large area is available. They require a long hydraulic retention time (HRT) and may produce bad odors, they have a slow treatment rate but can resist the shock loads. They can be used in low-income countries where funds are limited (Singh et al. 2015). Examples of this system include constructed wetlands (CWs), vermin-filters, horizontal gravel filters, multi soil layer systems, stabilization ponds, etc.

3.4 Aerobic Systems

The energy consumption of these systems is high but these can produce good quality effluent but requires semi-skilled personnel. The aerobic systems are high rate systems and are sometimes used as a post-treatment technique for the effluent of anaerobic units. These systems have a high biomass production but with good settling characteristics. (Singh et al. 2015). The typical examples of aerobic systems reported in the literature are activated sludge processes (ASP), moving bed reactors (MBR), gravity-driven membrane bioreactor, nitrification, and denitrification trickling filters.

3.5 Anaerobic Systems

These systems are low energy consuming, cheap and simple, and energy provider. Its effluent achieves only low to moderate effluent quality thus requires post-treatment, these processes take a long to start-up time and have odor nuisance (Singh et al. 2015). Anaerobic systems include UASBs, septic tanks, Baffled septic tanks, anaerobic baffled reactor.

3.6 Combination Aerobic and Anaerobic System

The combination of the system includes, an anaerobic tank and soil trench system; anoxic filters and multi-stair driving wheel rotating biological contactors; anaerobic bioreactor (ABR) and chemical catalytic bioreactor.

4 Summary of Decentralized Wastewater Treatment Systems

Various Researchers have proposed various decentralized wastewater treatment systems, employing various technologies. So, a summary of those systems has been provided in the Table 1.

5 Conclusions

From the current status of sewage generation, problems encountered in sewage treatment, negative environmental impacts of untreated sewage discharge as well as on the basis of literature survey, it is concluded that.

- Decentralized sewage treatment can prove to be a sustainable option or solution as an alternative to the centralized systems.
- There are various technologies used in decentralized systems reported in the literature which give different results under different operating conditions, so a proper choice of technology should be made, it can be based on the combination of various technologies under different operating conditions.
- No universal solution exists that which technology should be selected, there are varying degrees of sustainability that technology is selected.
- Only a few countries have done the economic analysis of these systems both in developed and developing countries, so the economic evaluation can be carried out for developing countries like India.
- Developing countries mostly employed anaerobic and combined treatment systems but the economic evaluation is almost non-existent.
- It can be seen that most of the studies are from developed countries but very few studies are from developing countries, so there is a need to carry the evaluation for the developing countries like India.
- A decentralized system can be a desirable and unavoidable approach to provide a locally appropriate solution.

Table 1 Summary of decentralized wastewater treatment systems

Type of system	% removal										Country	Reference
	COD	BOD	TSS	TN	TP	TDS	TC	NH_4^+–N	NO_3^-–N	PO_4–P		
MBR system	–	–	–	90	70	–	–	–	–	–	Switzerland	Abegglena et al. (2008)
Baffled septic tanks	74.8	71.4	71.1	–	–	30	–	–	–	–	Vietnam	Anh et al. (2002)
Anaerobic baffled reactor followed by horizontal gravel filter	72	84	61	–	–	–	–	–	–	–	India	Sutar and Kulkarni (2016)
Baffled anaerobic septic tank (BAST)	76	–	78	–	–	–	–	–	–	–	Vietnam	Nguyen et al. (2008)
Decentralized sewage treatment reactor	76	–	84.2	–	–	–	–	77.1	–	–	China	Dong et al. (2013)
Advanced septic reactor	86.2	79.4	95	–	–	–	–	–	–	–	Iran	Jorsaraei et al. (2014)
Combination of anaerobic tank and soil infiltration system	96	–	97	75	97	–	–	99	–	–	China	Ye et al. (2008)
Subsurface wastewater infiltration system	84.8	–	–	80.7	–	–	–	84.8	–	–	China	Ying et al. (2011)
Multi soil layer system	98.3	–	–	69.8	100	–	–	76.6	96.1	–	China	Honga et al. (2019)
Up-flow septic reactor	84	81	89	–	–	–	–	–	–	–	Egypt	Sabry (2010)
Gravity driven membrane bioreactor	–	–	–	–	–	–	–	–	–	–	China	Ding et al. (2017)
Multi-stair driving wheel rotating biological contactor	88.4	–	–	52.3	34.1	–	–	88.1	–	–	China	Xiao et al. (2018)

(continued)

Table 1 (continued)

Type of system	% removal										Country	Reference
	COD	BOD	TSS	TN	TP	TDS	TC	NH_4^+–N	NO_3^-–N	PO_4–P		
Chemical catalytic bioreactor	97	–	–	94	93.3	–	–	–	–	–	China	Zhifeng et al. (2020)
Vermi-filtration	90	90	95	–	–	92	–	–	–	–	Brisbane	Sinha et al. (2008)
Chemical biological adsorption followed by algal treatment	74	–	82	67	96	–	–	–	–	–	Belgium	Zamalloa et al. (2013)
On-site mobile septage treatment unit	80	–	81	–	–	–	99	–	–	–	India	Forbis-Stokes et al. (2021)
Hybrid electrochemical GAC reactor	69	–	71	–	–	–	–	–	–	–	USA	Rogers et al. (2018)
Trickling nitrification and denitrification filters	84	–	–	69	–	–	–	–	–	89	USA	Hunter and Deshusses (2019)
Trickling nitrification and denitrification filters	75	–	–	49	50	–	–	–	–	–	USA	Forbis-Stokes et al. (2020)

References

Abegglena C, Ospelta M, Siegrist H (2008) Biological nutrient removal in a small-scale MBR treating household wastewater. Water Res 42:338–346

Anh NV, Ha TD, Nhue TH, Heinss U, Morel A, Moura M, Schertenleib R (2002) Decentralized wastewater treatment—new concept and technologies for Vietnamese conditions. Int Water Assoc 33(7):1559–1578

Asano T, Levine AD (1996) Wastewater reclamation, recycling, and reuse: past, present, and future. Water Sci Technol 33:1–14

Boller M (1997) Small wastewater treatment plants—a challenge to wastewater engineers. Water Sci Technol 35:1–12

Bouwer H (1993a) From sewage to zero discharge. Euro Water Pollut Control 3:9–16

Bouwer H (1993b) Urban and agricultural competition for water and water reuse. Water Res Develop 9:13–25

Chawathe SD, Kantawala D (1987) Reuse of water in city planning. Water Supply 15:17–23

Chong MN, Ho ANM, Gardner T, Sharma AK, Hood B (2013) Assessing decentralized wastewater treatment technologies: correlating technology selection to system robustness, energy consumption, and GHG emission. J Water Clim Change 4(4):338–347

Dhote J, Ingole S, Chavhan A (2012) Review on wastewater treatment technologies. Int J Eng Res Technol 1:1–14

Ding A, Wang J, Lin D, Tang X, Cheng X, Guibai L, Nanqi R, Liang H (2017) In situ coagulation versus pre-coagulation for gravity-driven membrane bioreactor during decentralized sewage treatment: permeability stabilization, fouling layer formation and biological activity. Water Res 126:197–207

Dong SS, Bao CH, Hua JF, Bo Z, Jiang MZ, Hai YZ, Pei QL (2013) Performance of a novel decentralised sewage treatment reactor. J Chem 10:2314–2319

Fach S, Fuchs S (2010) Design and development of decentralized water and wastewater technologies: a combination of safe wastewater disposal and fertilizer production. Water Sci Technol 62(7):1580–1586

Forbis-Stokes AA, Miller GH, Segretain A, Rabarison F, Andriambololona T, Deshusses MA (2020) Nutrient removal from human fecal sludge digestate in full-scale biological filters. Chemosphere 257:127–219

Forbis-Stokes AA, Kalimuthu A, Ravindran J, Deshusses MA (2021) Technical evaluation and optimization of a mobile septate treatment unit. J Environ Manage 2:111–361

Frijns J, Jansen M (1996) Institutional requirements for appropriate wastewater treatment systems. In: Balkema A, Aalbers H, Heijndermans E (eds) Workshop on sustainable municipal waste water treatment systems, Leusdan, the Netherlands. ETC in cooperation with WASTE. 12–14 November 1996, pp 54–66

Ho G (2005) Technology for sustainability: the role of onsite, small and community scale technology. Water Sci Technol 51(10):15–20

Honga Y, Huang G, Anc C, Songa P, Xing X, Cheng X, Zhang P, Zhao Y, Zheng R (2019) Enhanced nitrogen removal in the treatment of rural domestic sewage using vertical-flow multi-soil-layering systems: experimental and modeling insights. J Environ Manage 240:273–284

Hunter B, Deshusses MA (2019) Resources recovery from high-strength human waste anaerobic digestate using simple nitrification and denitrification filters. Sci Total Environ 99:19–23

Jorsaraei A, Gogol M, Van Lier JB (2014) A cost-effective method for decentralized sewage treatment. Process Saf Environ Prot 9(2):815–821

Mara D, Cairncross S (1989) Guidelines for the safe use of wastewater and excreta in agriculture and aquaculture. United Nations Environmental Programme/World Health Organization, Geneva

Massoud MA, Tarhini A, Nasr JA (2009) Decentralized approaches to wastewater treatment and management: applicability in developing countries. J Environ Manage 90:652–659

Nguyen Anh V, Pham Nga T, Thang H, Morel A (2008) SANDEC: improved septic tank, a promising decentralized wastewater treatment alternative in Vietnam. Water Res 33(7):356–359

Rogers TW, Rogers TS, Stoner MH, Sellgren KL, Lynch BJ, Forbis-Stokes AA, Stoner BR, Hawkins BT (2018) A granular activated carbon/electrochemical hybrid system for onsite treatment and reuse of blackwater. Water Res 46:3705–3711

Sabry T (2010) Evaluation of decentralized treatment of sewage employing upflow septic tank/ baffled reactor (USBR) in developing countries. J Hazard Mater 174:500–505

Singh NK, Kazmi AA, Stark M (2015) A review on full-scale decentralized wastewater treatment systems: techno-economical approach. Water Sci Technol 71(4):468–478

Sinha RK, Bharambe G, Chaudhari U (2008) Wastewater treatment by vermifiltration with synchronous treatment of sludge by earthworms: a low-cost sustainable technology over conventional systems with potential for decentralization. Springer Sci Bus Media 48:125–132

Sutar SA, Kulkarni GS (2016) Development of decentralized wastewater treatment system for kitchen wastewater. Int J Innov Res Creat Technol 2:2454–5988

Xiao Z, Jun M, Xiwu L (2018) Performance of a coupling device combined energy-efficient rotating biological contactors with anoxic filter for low-strength rural wastewater treatment. J Clean Prod 196:1106–1115

Ye C, Hu ZB, Kong HN, Wang XZ, He SB (2008) A new soil infiltration technology for decentralized sewage treatment: two-stage anaerobic tank and soil trench system. Pedosphere 18(3):401–408

Ying HL, Hai BL, Tie HS, Xin W (2011) Effects of hydraulic loading rate on pollutants removal by a deep subsurface wastewater infiltration system. Ecol Eng 37:1425–1429

Yu H, Tay J, Wilson F (1997) A sustainable municipal wastewater treatment process for tropical and subtropical regions in developing countries. Water Sci Technol 35:191–198

Zamalloa C, Boon N, Verstraete W (2013) Decentralized two-stage sewage treatment by chemical–biological flocculation combined with microalgae biofilm for nutrient immobilization in a roof installed parallel plate reactor. Biores Technol 130:152–160

Zhifeng H, Desheng L, Yuanhui L, Chao Z, Detian G, Hong Y (2020) Performance of chemical catalytic biofilm technology for decentralized sewage treatment. Environ Eng Sci 10:1041–1052

Zhongxiang Z, Yi Q (1991) Water saving and wastewater reuse and recycle in China. Water Sci Technol 23:2135–2140

GPSR Compliance

The European Union's (EU) General Product Safety Regulation (GPSR) is a set of rules that requires consumer products to be safe and our obligations to ensure this.

If you have any concerns about our products, you can contact us on ProductSafety@springernature.com

In case Publisher is established outside the EU, the EU authorized representative is:

Springer Nature Customer Service Center GmbH
Europaplatz 3
69115 Heidelberg, Germany

Batch number: 10732090

Printed by Printforce, the Netherlands